LORD KELVIN

and the Age of the Earth

Kelvin

From a photograph by Annan. Glasgow. 1897

LORD KELVIN

and the Age of the Earth

JOE D. BURCHFIELD

With a new Afterword

The University of Chicago Press
Chicago and London

The University of Chicago Press, Chicago 60637
The University of Chicago Press, Ltd., London

© 1975, 1990 by Joe D. Burchfield
All rights reserved. Originally published 1975
University of Chicago Press edition 1990
Printed in the United States of America

98 97 96 95 94 93 92 91 90   5 4 3 2 1

Library of Congress Cataloging-in-Publication Data

Burchfield, Joe D.
   Lord Kelvin and the age of the earth / Joe D. Burchfield.
      p.   cm.
   Reprint. Originally published: New York : Science History
Publications, 1975.
   Includes bibliographical references.
   ISBN 0-226-08043-9 (pbk.)
   1. Earth—Age.   2. Kelvin, William Thomson, Baron,
1824–1907.   I. Title.
QE508.B84   1990
551.7—dc20   90-10964
                                                         CIP

∞ The paper used in this publication meets the minimum
requirements of the American National Standard for Information Sciences—Permanence of Paper for Printed
Library Materials, ANSI Z39.48-1984.

For Aline and Dan

Contents

Preface

It takes a long time to prepare a world for man, such a thing is not done in a day. Some of the great scientists, carefully ciphering the evidences furnished by geology, have arrived at the conviction that our world is prodigiously old, and they may be right, but Lord Kelvin is not of their opinion. He takes the cautious, conservative view, in order to be on the safe side, and feels sure it is not so old as they think. As Lord Kelvin is the highest authority in science now living, I think we must yield to him and accept his view. He does not concede that the world is more than a hundred million years old. He believes it is that old, but not older.

Mark Twain, *Letters from the Earth*

The concern with determining the age of the earth in years appeared in the west with the birth of Christianity, but it was not until the second half of the nineteenth century that this concern developed into the subject of a full-scale scientific investigation. Between 1860 and 1930, however, the accepted estimate of the earth's age evolved from the virtually unlimited expanse of time demanded by Charles Lyell and his followers, through a series of ever more restrictive limits imposed by nineteenth century physics and geology, to the vast but quantitatively definable chronologies derived from radioactive dating. During that cycle, the question of the earth's age stirred heated debate, stimulated important studies in physics, astronomy, geology, and biology, and served as the focus for a fascinating interplay of personalities and hypotheses.

In the study presented here, I have traced the evolution of the conception of geological time by describing the theories and principles involved in the more important methods of calculation. In the process, I have also attempted to illustrate the importance of a conscious concern with quantification in the development of a scientific discipline. And finally, by concentrating upon the continuity of Kelvin's influence and the interaction between physics and geology, I hope that I have provided some insight into the way in which individual and collective preconceptions, the deference to authority, and the interactions between disciplines and personalities affect the development of scientific ideas.

It has been necessary, of course, to impose several important restrictions on the scope of my investigations. For one thing, I have concentrated upon the physico-geological problem of establishing a quantitative limit for the probable age of the earth rather than upon the nearly parallel but more exclusively geological problem of determining the relative ages of strata. I have also concentrated upon the interaction between only two branches of science, physics and geology, during the whole period under discussion. Biology and astronomy have been included when appropriate, but only in the case of Darwinian evolution, where the question of time is critical, is this discussion at all extensive. Religion, which was often central to the earlier debates over the earth's age, played a relatively insignificant part in the later controversies and has been considered here only when it had a definite effect upon scientific thought. And finally I have dealt almost exclusively with British scientists during the period before 1890 and primarily with British and American scientists thereafter.

This last restriction, which was imposed neither arbitrarily nor merely for convenience, deserves some explanation. The number of scientists actively pursuing the problem of geological time was never very large, and during the second half of the nineteenth century it was confined almost exclusively to a small group of British investigators. Geologists on the continent made impressive progress in the study of crystallography, mineralogy, stratigraphy, paleontology, and geomorphology, but with very rare exceptions they made little attempt to establish a quantitative geochronology. Continental physicists seem to have ignored the problem altogether. Furthermore, although I must admit that my research in Continental scientific literature—primarily French and German—was considerably less thorough than in British and American journals, the few studies I did find relied heavily upon British sources and differed little from their conclusions. American geologists, too, took their measure of time from their British colleagues. With a vast, largely undefiled geological laboratory stretching before them, they devoted themselves to exploration and observation rather than to speculation and theory-building. There were exceptions, of course, but in general it was not until near the end of the century that American geologists began to take an active part in the search for the age of the earth.

It is difficult to say why the age of the earth should be so exclusively an Anglo-Saxon preoccupation. Friends, more learned in Continental geology than I, have offered some interesting speculations which as yet I have been unable to check adequately. I can, however, offer some admittedly speculative suggestions for why the question attained such prominence in England. In the first place, despite its growing professionalization, geology remained the British popular science *par excellence* throughout the nineteenth century. It consequently continued to attract the skill and efforts of amateurs like Kelvin, Huxley, and William Hopkins who were established masters in a variety of subjects and who could bring fresh points of view to bear on geological problems. Such diversity of knowledge was particularly valuable in the case of the age of the earth because, as it turned out, geological methods alone were not sufficient to establish an exact chronology. Another important factor was the character of British scientific organizations, particularly the geological societies and the British Association. These organizations provided extremely congenial forums for noted amateurs and for the interdisciplinary exchange of ideas. Such societies were not uniquely British, of course, but in the clubbish atmosphere of Victorian science they frequently attained their fullest development. Certainly in the case of the age of the earth, they served as the principal sounding boards for new ideas and the forums for discussion and debate.

One final word is perhaps in order to explain the inclusion of radioactive dating in what is primarily a study of Victorian science. The discovery of radioactivity undoubtedly marks a watershed in the investigation of geological time for it not only destroyed the basis for Kelvin's calculations, but supplied the tools for an entirely new approach to the problem. There are nonetheless valid reasons for considering the new methods as a continuation of the old. For one thing, Kelvin's influence remained strong throughout the first two decades of the twentieth century even among advocates of the new time scales revealed by radioactivity. And for another, the geologists themselves were most unwilling to abandon the techniques, hypotheses, and results that they had so laboriously compiled. Before the new time scale could be accepted, therefore, it had to be shown to be compatible with the record revealed by the crust of the earth itself. Thus to end my investigation with radioactivity's first challenge would leave the story half told, and indeed might serve to reinforce already existing misconceptions about the effect of the discovery of radioactivity upon geochronology. Furthermore, the debates generated by radioactive dating cannot be fully appreciated without an understanding of Kelvin's continued influence at the end of the nineteenth century. And finally, certain parallels exist in the study of the earth's age both before and after the discovery of radioactivity which will, I hope, give this study some significance beyond that accorded the simple investigation of an esoteric scientific problem. The role of preconceptions and authority in interpreting data, the validity of comprehensive theories based upon limited data, the influence of one science upon another, and the conflicts between them are, I believe, all subjects worthy of careful study.

I have accumulated many debts during the preparation of this book, more than I can ever properly acknowledge. There are several people and institutions, however, to whom I am especially indebted and to whom I offer a special thanks. William Coleman first read parts of this work as a dissertation at The Johns Hopkins University, and in the years since, he has continued to provide invaluable criticism and assistance as well as even more valuable encouragement. Rhoda Rappaport, Martin J. S. Rudwick, Alexander Ospovat, Leonard G. Wilson, and Paul A. Carter all read earlier versions of the manuscript and offered me the benefit of their knowledge and insights. I have been cheerfully assisted by the staffs of many libraries, particularly those of The Johns Hopkins University Library, the Northern Illinois University Library, the John Crerar Library, the Geology Library of the University of Illinois, the Library of Congress, and the University of Chicago Library. Countless others have helped through interlibrary loan. The Faculty and Board of Regents of Northern Illinois University have provided financial support through several faculty research grants. I am also particularly indebted to Miss Sue Reynolds and Mrs. Darla Woodward of the History Department staff at Northern Illinois University for their patient help in preparing the final drafts of the manuscript. And finally my greatest thanks are reserved for my wife, who sustained me both by her encouragement and her perceptive criticism, and for my children who bore with it all without really knowing why.

J. B.

I
Introduction

Early in 1869, some nine years after his now legendary encounter with Bishop Wilberforce, Thomas Henry Huxley found himself again the advocate for a beleaguered scientific theory. His task was to defend uniformitarian geology against the charge "that a great mistake has been made—that British popular geology at the present time is in direct opposition to the principles of Natural Philosophy."[1] This time, however, Huxley's opponent was not the shallowly eloquent bishop whose scientific expertise had long since ended with an undergraduate first in mathematics. This time his opponent was William Thomson (1824-1907), a scientist of international reputation who, at age forty-four, was at the height of his powers. And this time the outcome was to be quite different.

Better known today as Lord Kelvin, the title his scientific accomplishments were eventually to win for him, Thomson was widely regarded by his contemporaries as the greatest physicist of his age.[2] But physics was only the most prominent of his interests. Equally at home with both the abstract theories and the practical applications of science, he was an accomplished mathematician, a successful engineer, and a scientific theorist whose interests ranged over every branch of physics including physical astronomy and physical geology. Throughout his life, starting with his earliest undergraduate papers, Kelvin devoted odd moments of his time to the application of physical principles to the problems of geology. By the mid 1850s this interest had convinced him that the uniformitarianism espoused by Charles Lyell and his followers must be wrong, and for the next forty years he relentlessly attacked its scientific basis. Time and again he charged that uniformitarianism had ignored the established laws of physics and had thus led geology into serious error. He was determined to rectify that error.

Kelvin's initial efforts met with only limited success. He began his attack just as uniformitarianism reached the peak of its influence as the dominant geological theory in Britain. And because his arguments were couched in physical rather than geological terms, for several years they passed unnoticed by the majority of geologists. His frustration mounted until in 1868 he delivered a vigorous polemic asserting that "a great reform in geological speculation seems now to have become necessary" if geology were to be returned to the path of true science.[3] The challenge could no longer be ignored. Indeed, by 1868 it had become obvious that Kelvin's attack, particularly as it related to geological time, struck not only at uniformitarianism but at Charles Darwin's theory of natural selection as well. Huxley felt obliged to reply. On rising to the defense, however, he found that he could not substantively answer Kelvin's objections; he could only attempt to divorce biological evolution from the uniformitarian concept of time. Although he waded into the fray with his customary skill, elegance, and wit, even he recognized that they alone were no match for Kelvin's physics.[4] Thus in the end, his "defense" became but one of the first steps in a reform of geological thought that was little short of revolutionary.

Historians of geology are now generally agreed that Kelvin's influence on late Victorian geology was tremendous, but they are less than unanimous in their assessment of its effect. At a recent conference, for example, Leonard Wilson observed that "the influence of Lord Kelvin on geological thought over the past century remains profound and will not be easily erased."[5] Wilson considers this influence to have been detrimental and speaks of "the damage done to geology during the period from 1865 to 1903," because "a whole generation of geologists had grown up uncertain of their principles in theoretical geology."[6] According to Wilson, Kelvin was largely responsible for discrediting Lyell and obscuring the historical significance of the concept of uniformity. At the same conference, however, R. H. Dott, Jr. declared that Kelvin had been "basically correct in attacking the popular nineteenth-century Lyellian concept of uniformitarianism as amounting to perpetual motion—a physical absurdity." Rather than retarding geological theory, Dott argues that "in effect, Kelvin laid the ground for a modern, *evolutionary view* of the earth."[7] My own view falls somewhere between these extremes, although it probably comes closer to Dott's than to Wilson's. Kelvin's attack did moderate Lyell's influence and force a retreat upon the uniformitarians. Indeed, on the question of geological time he forced the reversal of the pendulum too far and pushed geologists into some treacherous and misleading speculations. He was no doubt also largely responsible for the over harsh criticisms which the geologists of the next generation leveled at Lyell and his followers.

But if Kelvin succeeded in modifying the doctrine of uniformitarianism, he did not abolish it. On the contrary, the belief in the uniform action of natural laws and geological processes continued to be a cardinal tenet of geological reasoning even among Kelvin's staunchest supporters. Uniformity alone was not enough, however, and the geologists of the second half of the century recognized the need to expand their horizons, to develop new methods of approach, to amass new and better data, and to utilize the discoveries of other sciences. Thus, thanks in great measure to Kelvin's influence, the quantitative study of geology came into its own. Moreover, this influence went far beyond the questions on which he worked directly—the formation of the earth, its age, and its internal constitution—and affected such diverse topics as paleontology, stratigraphy, geomorphology, orogeny, and even nuclear geology. And perhaps most important of all, it stimulated an interchange of ideas between physics and geology which benefited both.

Historical Background

The problem that concerned Kelvin was relatively new as scientific questions go. Although cosmogonies and theories of the earth's formation are as old as Egypt and Sumer, only rarely has man concerned himself with dating the time of that formation in years. Indeed, only in the last few centuries has he had any but the most vague conception of geological change. Consequently, when he speculated on the earth's beginning, he usually thought of it as being created more or less instantaneously, fully formed, and in nearly its present condition. Furthermore, the time of the beginning was always shrouded in myth. Even in those societies with a well-developed sense of historical time, the beginning of all things lay beyond history, and as Toulmin and Goodfield have pointed out, the transition between historical and mythical time was murky indeed.[8] Francis Haber put it more bluntly when he declared that, "Archaic societies revolt against concrete, historical time, and try to return periodically to a lost paradise of mythical time at the beginning of things."[9] Thus the ancient Egyptians envisaged vast but indeterminate ages for their past history, while eastern societies and many of the Greeks saw time as an endless succession of cycles. In short, the concept of geological time in the ancient world, if it existed at all, tended to be vague and qualitative. To date *The Beginning* was meaningless.[10]

The change came with the rise of Christianity. Christ could die for man only once, the early church fathers argued. The flow of time must therefore be unique and rectilinear, begun at the Creation and soon to be terminated at the Last Judgment. The more literal minded of the church fathers recognized no distinction between historical and mythical time. They viewed the genealogies of *Genesis* as a complete and accurate chronicle of the time since the first day of Creation. *The Beginning* could thus be dated accurately and exactly in years. Such calculations admittedly had been made by the Jews as early as the first century B.C., but it was the Christian apologists who first awakened Western man to a concern for the age of the world. The early apologists did not confine their efforts to *Genesis,* however. Both Theophilus of Antioch (c.115-c.183) and Eusebius of Caesarea (c.260-c.340), to take only the most notable examples, studied and compared the existing histories of the East before developing a Christian chronology. Still, probably the most influential of the Biblical chronologies was that of Julius Africanus (fl.200-250), who based his calculations upon the belief that all history must comprise a cosmic week with each day lasting a thousand years. But since Julius believed that Christ had come during the sixth "day" of the cosmic week, he placed the Creation at about 5500 B.C. (some accounts give 5002 B.C.), a figure in close agreement with the 5529 B.C. obtained by Theophilus. Numerous other chronologies were produced which, because of variations in the texts and commentaries used, differed slightly in the exact date of Creation; but all agreed that the antiquity of the earth was little greater than the history of man, and that they all confined to a span of five or six thousand years.[11]

The medieval church was undisturbed by the minor variations in the different chronologies. Its theology, after all, was not bound by Biblical literalism, and it retained the authority to interpret the scriptures. Haber suggests that even the eternalism of Aristotle may have returned to the West with the rise of Scholasticism. Literalism, however, returned at full force with the Reformation, and with it came new Mosaic chronologies undefiled by the profane histories of the East. The new chronologies, like Luther's (1483-1546) 4000 B.C. for the date of Creation, and Bishop Ussher's (1581-1656) 4004 B.C., were even more restrictive and exact than those of Theophilus and Julius. Indeed, there were those who proposed to set even the day and hour of Creation. Bibliolatry reigned throughout the sixteenth and seventeenth centuries, and Mosaic chronology went virtually unchallenged. The question to be answered by the natural philosopher was not how long it had taken for geological forces to sculpture the face of the earth, but how had they accomplished their work in the few millennia since Creation. In one way or another, the cosmogonies of Leibniz (1646-1716) and Descartes (1596-1650) and the "theories of the earth" of Thomas Burnet (1635-1715), John Woodward (1665-1728) and William Whiston (1667-1752) were all directed toward this end. In the process, however, they introduced a dichotomy that was profoundly to influence the study of the age of the earth, the rival claims of fire and water as the primary agent of geological change.[12]

The unchallenged dominance of Mosaic chronology ended with the Enlightenment. Dates of Creation based upon *Genesis* still prevailed in theological and popular literature, but the desire of philosophers to find natural causes in a Newtonian universe led inexorably away from a world created fully-formed in an instant of time. Before the end of the eighteenth century, both Immanuel Kant (1724-1793) and Pierre Laplace (1749-1847) had independently formulated the famous nebular hypothesis in which the Newtonian laws of gravitational attraction were extrapolated to explain the gradual evolution of the solar system. During the same years another Newtonian, the Comte de Buffon (1707-1788), abandoned Biblical chronology and attempted to determine the earth's age by experiment and scientific reasoning. Buffon's approach is especially interesting since it was based upon the evidence of the earth's internal heat and estimates of its rate of cooling—the same fundamental idea, although quite different in specific detail, that Kelvin was to exploit so fully a century later. Furthermore, by placing the time of Creation some 75,000 years in the past, Buffon broke sharply with historical time and thus liberated the age of the earth from the restrictions of the age of man.[13]

It was not the Newtonian mechanical universe that impressed the need for time upon Enlightenment natural philosophy, however. It was the rise of paleontology. From the time of the Greeks, fossils had stirred men's imaginations and had given rise to some remarkable speculations. They had played their part in the seventeenth century theories of the earth; and both Robert Hooke (1635-1703) and Nicolaus Steno (1638-1687) had advanced extraordinarily perceptive accounts of their origin. But it was not until the eighteenth century that the study of fossils began in earnest, and it soon fostered a growing conviction that the face of the earth had changed many times since its

creation. Time seemed a necessary corollary to such a conclusion, time
to accommodate the succession of living forms. Indeed, for some more
daring or more speculative thinkers this process of change seemed
endless. Benoît de Maillet (1656-1738), for example, revived the cyclic
eternalism of the East and denied that there had been a *Beginning* at
all.[14] Others, such as James Hutton, accepted a beginning but denied
the possibility of finding records of it in the earth itself. Perhaps the
most conclusive argument, however, came from the great rival paleontologists Georges Cuvier (1769-1832) and Jean-Baptiste Lamarck
(1744-1829). As different as were their views on geology and the
nature of life, Cuvier and Lamarck both agreed that the succession of
forms revealed by the paleontological record required far more time
than the chronologies of *Genesis* would allow. The orthodox Cuvier
might accept the reality of the Noachian Deluge some five or six
thousand years ago, but he had to extend antediluvian time much
farther into the past, perhaps even indefinitely. Lamarck, too, regarded water as the principle agent of geological change, but he saw no
reason to limit its action by the account of Noah or any other chronicle.
History, Lamarck argued, was no guide to the earth's antiquity. The
earth's surface was its own historian, and the only reliable chronicle was
the record of the origin and gradual development of living organisms.
Even that record, however, could not reveal the earth's age in years,
and faced with it Lamarck could only exclaim: "Oh how very ancient
the earth is! And how ridiculously small the ideas of those who consider
the earth's age to be 6000 odd years."[15]

In spite of the daring speculations of Buffon and Lamarck, however,
it was Britain rather than France which was to nurture the study of
geological time. After Hooke, Woodward, and Whiston, British geology suffered a general hiatus for more than three-quarters of a century. Then in 1785, seven years after the appearance of Buffon's
Epochs of Nature and seventeen years before Lamarck's *Hydrogeologie,*
James Hutton (1726-1797), a Scottish gentleman-farmer and physi-
cian, published a paper that was to inaugurate a revolution in
geology.[16] Years spent exploring the varied geological formations of
England and Scotland had convinced Hutton that universal deluges
were not part of the economy of nature. Indeed, except for Creation
itself, he saw no need to invoke catastrophes of any kind; the normal
agents of geological change could, if given sufficient time, account for
all the changes which have shaped the earth's surface. It was obvious
that Hutton's conception demanded vast ages for the slow destruction

and rebuilding of the continents, but like the theorists of the seventeenth century, his concern was not with time but with the agents of change. From the dichotomy of fire and water, he sought to assert the dominion of fire; time was of secondary importance. For Hutton, the geological record clearly proclaimed a cyclic progression of changes so ancient as to obscure any "vestige of a beginning," and to hold out "no prospect of an end."[17] These phrases did not deny a beginning or an end, as they were often misinterpreted to imply; they asserted only that time was so incomprehensibly vast as to defy calculation. Its exact duration was therefore of little consequence.[18]

Hutton's opponents did not share his indifference to exact chronology. For one thing, the so called "golden age of geology," which his work is often credited with inaugurating, coincided with a period of conservative reaction in religion and an accompanying return to bibliolatry and strict scriptural literalism. This reaction, due perhaps in part to a revulsion against the excessive rationalism of the French Revolution, produced an atmosphere of piety which was highly congenial to the revival of Mosaic chronology. The long eons of time postulated by the eighteenth century rationalists became just one more manifestation of the threat that rationalism posed to religion and proper order. Richard Kirwan (1733-1812), the respected phlogiston chemist and Hutton's implacable opponent, summed up this opinion as succinctly as any "by observing how fatal the suspicion of the high antiquity of the globe has been to the credit of Mosaic history, and consequently to religion and morality."[19]

Kirwan's opposition to Hutton and that of many of his fellows was based on more than religion, however. As a spokesman for the Neptunist school of geology (the advocates of water as the principle agent of geological change), he opposed Hutton's advocacy of the primary role of heat. Still, religion, Neptunism, and the question of time were not unrelated. If the long eons demanded by Hutton and other eighteenth century rationalists were to be denied, periodic catastrophes seemed necessary to account for the tremendous changes that had shaped the earth's surface. And periodic large scale floods could provide such catastrophes. To be sure, Neptunism did not in itself impose a necessary limit on time. In fact, Neptunism included among its more notable proponents several champions of the earth's great antiquity, including Lamarck and De Maillet; while the advocates of heat (called Vulcanists or Plutonists) included some, like Hutton's friend James Hall (1761-1832), who would compress the demand for time by invoking their own version of plutonic catastrophes. In general, however, Neptunism was the favored doctrine of those who sought to hurry the pace of geology and to vindicate the chronology of Moses.

In addition to its unquestioned appeal to Biblical literalism, it should be stressed that Neptunism had strong scientific evidence in its favor as well. Both the stratigraphic record and the distribution of marine fossils provided impressive empirical evidence for the action of water; and before the discovery of recurring epochs of widespread glaciation, many of the features now attributed to the work of ice appeared to be explicable only in terms of the action of flood waters.[20] Furthermore, the scientific aspects of Neptunism were forcefully expounded by the most influential teacher of geology of the age, Abraham Gottlob Werner (1749-1817), and at a time when the influence of the German university tradition was rising rapidly in England, Werner's ideas enjoyed a cordial reception. It is not surprising, therefore, that this seemingly irrefutable alliance between orthodox science and religion should effectively eclipse Hutton's influence for several decades.

Nonetheless, in the flurry of geological activity that greeted the nineteenth century, it became increasingly difficult for even the most ardent catastrophist to compress the earth's history into a few thousand years. Although it still seemed possible that all the geological changes attributed to water could have taken place during the forty days of the Deluge, such a conclusion grew increasingly less probable or even theoretically desirable. Thus, even Werner, whose own chronology was frankly confusing, admitted late in life that the earth had built up gradually over time. Similarly, William Buckland (1784-1856), Werner's British counterpart both as the leader of the Neptunists and as the teacher of a whole generation of geologists, found it necessary to assume a long period of geological development preceding the deluge of Noah. As a result, various ideas about "extended days of creation," ideas very similar to those of Cuvier, came again into vogue. Haber has admirably characterized the subsequent attempt at *ad hoc* compromise:

> The chronological outlook during the 1820's may be summed up by saying that the traditional Biblical view prevailed for the most part, but among geologists and harmonists a division was made between antediluvian and postdiluvian times, supernatural forces prevailing in the former and only natural forces in the latter. By confining Biblical chronology to man's history alone, they claimed a measure of freedom in the antediluvian

period to fill out the segments of geochronology on the
basis of observed evidence, but the approach was
piecemeal and without any over-all estimate of the
duration of the earth, except that it was "vast."[21]

Despite such attempts at compromise, the spectre of Moses continued to hang over geological chronology for several decades. In
Huxley's acerbic metaphor, the path of geological speculation was
blocked by a thorny barrier carrying the notice: "No Thoroughfare. By
order Moses."[22] Hutton had broken a hole in the barrier, and in the
1820s it was significantly enlarged by George Poulett Scrope
(1797-1876), but the major breach came in 1830 with the publication of
Charles Lyell's (1797-1875) *Principles of Geology*.[23]

Historians, influenced by the polemic cast of Lyell's writings and
those of his contemporaries, have tended to emphasize his insistence
upon *actualism*—the use of the present as the only valid guide to the
past—and his opposition to the prevailing reliance upon catastrophes.
Recently, however, Martin J. S. Rudwick has suggested that the real key
to Lyell's view of nature "was his extension of the time scale of earth
history; or rather, his imaginative grasp of the full implications of such
a time scale for the interpretation of geology."[24] If the vast reserves of
force demanded by the catastrophists were to be rejected, they had to
be replaced; and like Hutton, Lyell's substitute was time. In Lyell's
view, the earth exists in a state of dynamic balance. Except for occasional, purely local variations, the forces acting on its surface were
assumed to have remained constant both in kind and degree, through
interminable ages. Again like Hutton, he did not view time as infinite,
but indefinite or inconceivably vast. Lyell's advantage over Hutton was
that he did not attempt another "theory of the earth," but produced
instead a careful analysis of the most recent of the great geological eras.
He was therefore able to demonstrate, far more effectively than Hutton, the efficacy of existing geological forces if only given time, and
thus to refute the need for catastrophes. Lyell's success was far from
instantaneous. He unquestionably shook the authority of the di-
luvialists, who were to be dealt still another blow a few years later by
Louis Agassiz's glaciers, but catastrophism in a more sophisticated
form remained a viable doctrine for decades. Indeed, it was not until
after 1850 that Uniformitarianism (a label coined by the catastrophist
William Whewell) could accurately be called the dominant geological
doctrine in Britain, and even then its dominion was never
unchallenged.[25]

In the polemical atmosphere surrounding the emergence of uniformitarianism, Lyell's followers sometimes abandoned his more cautious choice of phrase. With regard to time, the word *infinite* frequently took the place of his more careful *indefinite* or *inconceivably vast*. Time became the uniformitarians' panacea. They drew on it lavishly but failed to grasp its full significance. *Inconceivably vast* was consequently more than a phrase; it was an accurate description of the limits of comprehension. This sense of vagueness was no doubt intensified by Lyell's rejection of progressionism in both geology and biology. His steady-state conception of the earth, after all, provided no real reference points from which to measure time. Thus, it was Charles Darwin (1809-1882), rather than Lyell, who first gave the uniformitarian conception of time a sense of perspective. Darwin was Lyell's disciple except on the question of biological progression, but it was that difference which gave time a direction.[26] As Archibald Geikie observed:

> Until Darwin took up the question, the necessity for vast periods of time, in order to explain the characters of the geological record, was very inadequately comprehended. Of course, in a general sense, the great antiquity of the crust of the earth was everywhere admitted. But no one before his day had perceived how enormous must have been the periods required for the deposition of even some thin continuous groups of strata. He supplied the criterion by which, to some degree, the relative duration of formations might perhaps be appointed.[27]

Admittedly, Darwin's one attempt at calculating geological time was a devastating fiasco.[28] But nothing could obscure the significance of his progressionism, both in what it revealed immediately about the geological record and in its significance for Lyell's steady-state world view.

In the years following the publication of *On the Origin of Species,* however, the central figure in the search for an accurate measure of geological time—or more exactly, the age of the earth—was to be neither Darwin nor Lyell, but Kelvin. And the central premises upon which the case was to be decided were drawn from physics rather than from biology or geology. Darwin had helped to create an intellectual atmosphere in which geologists were just becoming fully conscious of the implications of time, but in which vague allusions to indefinite or infinite ages still prevailed. Kelvin injected into this atmosphere a reinvigorated conception of time as exact, finite, and brief; a conception, moreover, based upon a belief in the uniformity of natural causes, not upon the dictates of religion. Kelvin rejected Darwinian evolution because it neglected the evidence of design and order in nature, but he rejected uniformitarianism because it neglected the laws of physics. Repeatedly for nearly forty years, Kelvin attacked the scientific basis of uniformitarianism, and his influence was such that he succeeded in changing the very meaning of the term. Indeed, in this regard, he did more than merely obscure the historical significance of Lyell's work, as Leonard Wilson has justly charged; he perverted it. For more than a generation Lyell's conception of geological time and, more particularly, those of his early followers were looked upon as the irresponsible demands of vaguely unscientific theorists.

Far from being a mere voice of opposition, however, Kelvin's accomplishment was to dispel the mists surrounding the question of the earth's age and focus attention upon the problem of establishing an exact geochronology. Almost single-handed he overthrew the mid-century uniformitarians' demands for unlimited time, and in the process stimulated a growing awareness of the necessity for quantitative measurement in geology. The generation of geologists who rejected the excesses of Lyell and his followers eventually came to reject Kelvin's restrictions as well; but at the century's end the overwhelming majority, whatever their specific disagreement with Kelvin might be, felt that they owed him a tremendous debt. The essence of this feeling was summed up in 1897 by Thomas Chrowder Chamberlin (1843-1928), one of America's most influential geologists. Although sharply critical of Kelvin's later pronouncements on geological time, Chamberlin had only praise for his overall influence:

> ... however inevitable must have been the ultimate recognition of limitations [of time] it remains to be frankly and gratefully acknowledged that the contributions of Lord Kelvin, based on physical data, have been most powerful influences in hastening and guiding the reaction against the extravagant time postulates of some of the earlier geologists. With little doubt, these contributions have been the most potent agency of the last three decades in restraining the reckless drafts on the bank of time. Geology owes immeasurable obligation to this eminent physicist for the deep interest he has taken in its problems and for the profound impulse which his masterly criticisms

have given to broader and sounder modes of inquiry.[29]

Measuring the Age of the Earth

Once the significance of determining the age of the earth had been
revealed, Kelvin's contemporaries responded by developing a remarkable variety of techniques for approaching the problem. Geologists,
physicists, biologists, and astronomers all contributed to the investigation, and their data and methods of approach reflected the diversity of
their interests. Ultimately, however, the problem always reduced to
one of finding a reliable geological or cosmic clock. Some phenomenon
had to be identified which could be readily quantified and measured,
and which could be shown to vary regularly with time. Today the decay
of radioactive elements is universally regarded as the only phenomenon which meets these criteria. In the nineteenth century the candi-
dates were much more numerous, although in general they can be
separated into three broad categories: physical chronometers, which
usually depended upon the laws of Newtonian physics and thermodynamics; astronomical chronometers, which related terrestrial
history to some cyclic or secular change in the solar system; and geological chronometers, which utilized the records left in the crust of the
earth itself. From time to time, biological clocks based on the evolution
of living forms were also introduced; but because of the difficulty of
measuring the rate of evolutionary change, such chronometers were
generally conceded to be valid only as measures of relative time.

Like nearly all attempts at categorization, the scheme proposed
above is at best only approximate, and many calculations of the earth's
age were proposed which borrowed elements from more than one
category. One thing that all of the methods had in common, however,
was the fact that only a few of the parameters necessary for calculation
were susceptible to direct measurement. The others had to be estimated either by analogy, indirect calculation, or simply educated
guessing. To be sure, the scientists involved worked hard to expand
and improve their store of data and to refine their calculations. The
geologists in particular became inspired with a zeal for quantification.
Throughout the nineteenth century they carefully measured, mapped, and classified the layers of accumulated strata; they explored the
ocean bottoms and the continental shelves; and they measured the
rates and intensities of the agents of geological change. But it was never
enough. In the end, geologists, physicists, and astronomers alike were

forced to estimate vital parameters that defied exact measurement; and thus inevitably they introduced the unquantifiable influences of their individual preconceptions and their reliance upon accepted authority.

Looking first at the physical approaches where Kelvin's influence was most direct, the principal arguments deal with the problem of energy, or more exactly, with the problems of production and supply of heat. Here, the fundamental principle is deceptively simple. Since long before history men have observed that the forcible collision of two bodies generates heat. This principle, which has been used to start innumerable campfires, was fundamental to the eighteenth century nebular cosmogonies of Kant and Laplace, but it was not until the nineteenth century that this most obvious phenomenon was rendered quantifiable. Due largely to the efforts of Julius Robert Mayer (1814-1878) and James Prescott Joule (1818-1889), the quantitative equivalence between mechanical work and heat was firmly established in the 1840's.[30] For the first time, the exact amount of heat that could be generated by a given amount of mechanical energy could be determined. A few years later, still another principle governing the properties of heat and energy was discovered by Kelvin and R. J. E. Clausius (1822-1888).[31] This, the second law of thermodynamics, states in its simplest terms that in any conversion of energy from one form to another, a proportion of that energy will be converted irretrievably into heat. In other words, in any closed system, including the universe as a whole, the total amount of energy will remain constant, but every energy transformation—that is, every physical process—will result in a diminution of the net amount of energy available for useful work. The universe must therefore be running down.

The laws of thermodynamics were to have far reaching scientific, philosophical, and literary effects during the second half of the nineteenth century. Initially, however, few geologists grasped their significance, and it was this failure that prompted Kelvin's incursion into the problem of geological time. Convinced by the principle of the dissipation of energy (the second law) that geological uniformity could not be a law of nature, he saw in the principle of the conservation of energy (the first law) the means by which to determine the limits of the ages of the earth and sun. Both of Kelvin's major arguments, and the subsequent revisions proposed by his followers, relied basically upon calculating the amount of energy available to the earth and sun from all conceivable sources; and then from measurements (or estimates) of the rates at which that energy is dissipated, determining the time necessary for each body to cool from its assumed primitive state to its present condition. The two cases naturally required different methods of analysis since the sun was assumed to be fluid and the earth solid; but in their fundamental premises, namely that the original condition of matter was a primitive nebula and that the only adequate conceivable source of energy was gravitation, both problems were impressively uniform.

Like the physical arguments, those drawn from astronomy had their roots in the dynamics of gravitational attraction. Throughout the eighteenth and nineteenth centuries astronomers had noted numerous variations in the motions of the solar system that could best be explained in terms of the gravitational attraction that every body in the system exerts upon every other body. The development of powerful techniques of mathematical analysis made it possible to calculate the past and future effects of these perturbations upon the orientation of the solar system, and as a result, two of them came to play significant roles in the search for an accurate measure of the earth's age.

The perturbations that most attracted the interest of Kelvin, his followers, and his opponents were the apparent secular acceleration of the moon's velocity with respect to the earth's period of rotation, and the periodic change in the ellipticity of the earth's orbit. In the former case, it was reasoned that if the moon were indeed accelerating and thus moving gradually away from the earth's surface, then it might be possible to calculate the time since its probable origin as a fragment thrown off from a rapidly spinning, still molten, primordial earth. Such a calculation, admittedly, would not give an exact measure of the earth's age, but it would supply an upper limit to the time during which the terrestrial crust could have been solid. Moreover, a corollary to this problem raised the possibility that the tides generated by the moon might act through frictional drag to slow the earth's velocity of rotation; and if so, it went on to ask whether the measure of such an effect might be used to determine how long the tides have been in operation. The problems of lunar origin and tidal drag both involved assumptions about the mode of origin, initial condition, and internal constitution of the earth and moon that were of great interest to physicists and astronomers, but neither was directly related to the concerns of geologists. In their view, the probable effects of changes in the eccentricity of the earth's orbit seemed much more directly relevant. Starting from the premise that any change in the earth's orbital eccentricity would necessarily alter the relative lengths of the seasons (when the earth's orbit is nearly circular as it is at present, for example, the seasons in the temperate zones are nearly equal in length; but during periods of pronounced eccentricity, the season experienced at the aphelion of the orbit will be much longer than at the perihelion), the question arose, could this periodic change in orbital shape be related to periodic changes in terrestrial climates, and in particular to the periodic occurrence of the glacial epochs? The question, of course, defied direct solution. But once the probability of an affirmative answer was established, two further questions were immediately raised: could astronomical calculations of past periods of high orbital eccentricity be used to date the glacial epochs, and if so, could those dates be used as standards for determining the durations of even more remote geological epochs? Between 1860 and 1900 affirmative answers were suggested for all of these questions, and depending upon the quantitative data employed, a variety of influential chronologies were developed.

Each of the astronomical, or astronomical-geological, approaches to time had much to recommend it, but the difficulties were obvious. In each case the impossibility of obtaining empirical data for many essential parameters placed a premium upon rigorous deduction and upon the feasibility of the assumptions involved. In the case of the lunar problem especially, the whole argument involved highly sophisticated mathematics which, although of great interest to astronomers and mathematicians, was beyond the scope of most geologists. Nonetheless, the basic hypotheses were extremely plausible, they were astronomically significant in their own right; and they could not be ignored.

As one might expect, the geological approaches to the question of time were based primarily upon the record revealed by the crust of the earth itself. That record had been carefully examined by the geologists of the first half of the nineteenth century; and from it, they had developed a scheme for dividing the earth's history into periods according to the forms of life that it has supported. The mere static record of the succession of life could provide no clue to how long the process of change had taken, however, and by mid-century it seemed apparent that the question of absolute time could be answered only by determining the rate at which the successive layers of strata had been deposited. The geologists of the second half-century turned to this task with a will, and the quantitative data needed for such calculations increased tremendously. The problem now lay in how this data was to be interpreted and in how the many questions that still could not be answered directly were to be treated. Not surprisingly, the result was a proliferation of methods and interpretations (some, admittedly, representing mere variations on a theme), and a great diversity of conclusions concerning the earth's age.

A major problem lay in the fact that the layers of strata, like the fossils they contain, present only a static record of past events. They provide, at best, only a relative measure of the rates at which they have been deposited, and even the rate of the present strata formation is concealed beneath the depths of the seas. The obvious alternative for geologists in the late nineteenth century was to grasp the stick by the other end. Sediment must be built up from the wreck of existing formations. Therefore, one of the geologists' first tasks was to determine the rate at which existing continents are worn away, and thus to find the rate of production of raw materials for sedimentation. By mid-century this task was already well underway as the areas of the world's major river basins and the sedimentary contents of the rivers were meticulously measured, analyzed, and compared. And by the seventies a high degree of agreement had been reached concerning the average rate of continental denudation. The task of the geological chronologist was to turn these data around and use them as a measure of the time required for the formation of existing strata. Such an approach required numerous assumptions, of course, including the most important that the rates of erosion and sedimentation have been nearly uniform throughout the earth's history, and that the present, for which data is available, represents a reasonable mean. Neither assumption presented any real difficulty during the late nineteenth century, since among geologists at least, Lyell's *actualism* continued to be accepted even after other parts of the theory of uniformitarianism had given way.

Even allowing for a high degree of uniformity in the past rates of geological activity, however, the geologists were still faced with the problem of determining the quantitative relationship between the bulk of sediment generated by erosion and the rate at which it is deposited to form new strata. Could he assume that the sediment was distributed uniformly over an area equal to the area denuded; or must the whole ocean floor be considered as a sedimentary basin; or, indeed, must he locate some small area along the continental shelves where all sedimentation must be assumed to be confined? Depending upon the answer to this question alone, the geologists' numerical results could and did fluctuate wildly, while other related questions added still more to the uncertainty of any conclusion. Among the questions for which no direct empirical answer was possible were: what relative importance should be assigned to sediments derived from mechanical and chemical denudation; what allowances should be made for unconformities and discontinuities in the geological record; and, indeed, could a general average be assumed for all kinds of rock, or did the rate of sedimentation have to be determined separately for each formation? And finally, there was the question of what correspondence to draw between the quantitative record and the relative ages provided by the older studies of fossils.

The task of sorting the alternatives was formidable, and the late nineteenth century produced advocates for virtually every possible permutation and combination. Estimates of time drawn from geology, astronomy, and physics were compared, revised, and compared again. Given the circumstances, the varied array of numerical results thus obtained is hardly surprising. It is only surprising that the range of variation was so limited. After 1870, only a handful of investigators defended a terrestrial age that differed much from Kelvin's 100 million years. A few would reduce his limit to the order of tens of millions, and a few others would increase it to hundreds of millions, but almost no one thought it necessary to exceed his upper limit of 500 million years. Even within that narrow range, however, the opportunities for controversy were endless, and some degree of tension characterized almost every discussion of geological time.

With the new century, the discovery of radioactivity introduced a group of new and entirely unexpected considerations into the search for the earth's age. In the first place, radioactivity revealed the existence of enormous stores of energy which threatened to undermine the basic premises of the old physical calculations. Certainly there was sufficient energy available to the earth and sun to expand the limits of time set by Kelvin and his followers many times over. Secondly, the transmutation of radioactive elements into entirely different elements provided investigators with an entirely new kind of geological clock. As was the case with many of the older methods of measuring time, however, the fundamentals of radioactive dating were relatively simple while the practical problems involved were formidable indeed. In principle, if one could determine the rate at which a particular radioactive element is transformed into an identifiable stable end product, the ratio of the parent and product elements should be proportional to the age of the minerals in which they are found. In practice, the tasks of definitely identifying the end products of radioactive decay and of determining the rates of transmutation proved to be enormous. Even the task of identifying mineral samples which were suitable for analysis required years of trial and error investigation. Thus in its early stages, radioactive dating, like the methods that it ultimately superseded, involved as many assumptions as data; and like them, it was subject to interpretation and open to dispute. Indeed, it was fully thirty years before the last of the Victorian chronologies gave way before radioactivity, and even then, they yielded only when the last participant in the earlier investigations was dead.

NOTE ON CITATIONS

In the notes at the end of each chapter, all books and articles are cited briefly by author, date, and abbreviated title. Page references are given where appropriate and volume numbers (in Roman capitals) for multi-volume works other than periodicals. Full bibliographic references for each citation may be found in the bibliography at the end of the volume, where they are listed by author, and under each author, by date of publication. Multiple publications by a given author in the same year are differentiated by small roman letters a, b, c, etc. In several cases (Kelvin's works, for example) page references have been made to collected works. Both the location of these references and the original place of publication have been included in the bibliography. For further information, see the explanatory note at the beginning of the bibliography.

REFERENCES

[1] Kelvin (1871a), *Geological Time*, p. 44

[2] William Thomson was not raised to the peerage as Baron Kelvin until 1892. In fact, he was yet to be knighted when he began his attack on uniformitarian geology although his scientific reputation was already well established. He is best known today, however, as Lord Kelvin, and so I have chosen to refer to him as "Kelvin" throughout this work except in direct quotations or where ambiguity might arise. The standard account of his life is still Thompson, S. P. (1910), *Kelvin*.

[3] Kelvin (1871a), *Geological Time*, p. 10.

[4] Huxley, T. H. (1869), *Geological Reform*, pp. 308-42.

[5] Wilson, L. G. (1969), *Intellectual Background*, p. 430.

[6] *Ibid.*, p. 427.

[7] Dott (1969), *James Hutton*, p. 140.

[8] Toulmin and Goodfield (1966), *Time*, p. 20.

[9] Haber (1959a), *Age of the World*, p. 11. Much of the material included in this brief

introductory background sketch is based upon the work of Haber and that of Toulmin and Goodfield. I am also indebted to L. Eiseley (1961), *Darwin's Century*; J. C. Greene (1961), *Death of Adam*; C. C. Gillispie (1959), *Genesis and Geology*; R. Hooykaas (1963), *Principle of Uniformity*; G. L. Davies (1969), *Earth in Decay*; and numerous papers by M. J. S. Rudwick and L. G. Wilson. Frequently the information that I have tried to compress into a few paragraphs cannot be attributed to a single source, and certainly none of the authors cited can be blamed for any oversimplifications that may have resulted.

[10]Haber (1959a), *Age of the World*, pp. 11-16, 38-44; Toulmin and Goodfield (1966), *Time*, pp. 55-64.

[11]Haber (1959a), *Age of the World*, p. 1, pp. 15-27; Toulmin and Goodfield (1966), *Time*, pp. 55-64.

[12]Haber (1959a), *Age of the World*, pp. 28-35, 44-98; Toulmin and Goodfield (1966), *Time*, pp. 64-95; Greene (1961), *Death of Adam*, pp. 25-63; Davies (1969), *Earth in Decay*, pp. 27-94.

[13]Haber (1959a), *Age of the World*, pp. 115-136, 146-159; Toulmin and Goodfield (1966), *Time*, pp. 129-135, 142-149; Meyer, H. (1951), *Age of the World*.

[14]DeMaillet's neglected work is now available in an excellent translation by A. V. Carozzi. See DeMaillet's (1969), *Telliamed*.

[15]Lamarck (1964), *Hydrogeology*, p. 75. Also see Haber (1959a), *Age of the World*, pp. 108-112, 174-179, 194-210.

[16]Hutton (1788), *Theory of the Earth*.

[17]*Ibid.*, p. 304.

[18]The best account of Hutton's ideas is still to be found in Playfair (1802), *Illustrations of the Huttonian Theory*. For briefer summaries, see Haber (1959a), *Age of the World*, pp. 164-173; Greene (1961), *Death of Adam*, pp. 84-93; Davies (1969), *Earth in Decay*, pp. 154-199 and others.

[19]Quoted in Gillispie (1959), *Genesis*, p. 55. Further discussion in Gillispie, pp. 20-72; Davies (1969), *Earth in Decay*, pp. 129-145; and Haber (1959a), *Age of the World*, pp. 7-8, 169-171, 191-194.

[20]Both historians and geologists have tended to minimize the scientific aspects of Neptunism. Gillispie, for example, implies that the Neptunist position was somehow *unscientific* (see especially: Gillispie (1959), *Genesis*, pp. 44-72). Such a view appears to me untenable. Certainly there were Neptunists, particularly in Britain, whose religion was stronger than their science, but their occasional foibles hardly nullify the mass of careful observation undertaken in the cause of the Deluge and in the cause of science. Wilson gives what seems to me a more balanced view of the Neptunists as scientists (Wilson, L. G. (1967), *Origins of Charles Lyell's Uniformitarianism*). Both M. J. S. Rudwick and R. Rappaport have pointed out the need for a more thorough study of the Neptunist position (Rudwick (1969), *Glacial Theory*, pp. 136-57 and Rappaport (1964), *Problems and Sources*, pp. 60-77).

[21]Haber (1959a), *Age of the World*, p. 214. For further discussion see Haber, pp. 187-215 and Gillispie (1959), *Genesis*, pp. 40-72, 98-120.

[22]Huxley (1897b), *Christian Tradition*, p. viii.

[23]Scrope (1827), *Geology of Central France*; Lyell (1830-33), *Principles of Geology*.

[24]Rudwick (1969), *Lyell on Etna,* p. 288.

[25]*Ibid.*, pp. 288-304; Rudwick (1970), *Strategy of Lyell's Principles,* pp. 5-33; Wilson, L. G. (1969), *Intellectual Background,* pp. 426-443; Cannon (1960), *Uniformitarian-Catastrophist Debate,* pp. 38-55; Gillispie (1959), *Genesis,* pp. 121-148.

[26]Rudwick has pointed out that the idea of progression or direction was very much alive during the early nineteenth century (Rudwick (1971), *Uniformity and Progression,* pp. 209-227). The rise of uniformitarianism seems to have tempered its influence, however, until the appearance of the *Origin.*

[27]Geikie, A. (1905a), *Founders,* p. 439.

[28]Darwin, C. (1859), *Origin,* pp. 282-287. A discussion of Darwin's treatment of time is given in chapter 3.

[29]Chamberlin, T. C. (1899), *Lord Kelvin's Address,* p. 890.

[30]Joule (1847), *Mechanical Equivalent of Heat,* pp. 173-176; Mayer, J. R. (1863b), *Mechanical Equivalent of Heat,* pp. 493-522.

[31]Clausius (1854), *Der mechanischen Warmtheorie,* pp. 473-555; Kelvin (1852), *Dissipation of Mechanical Energy,* pp. 511-514.

II
Kelvin and
the Physics of Time

The study of geology has always attracted and profited from the efforts of dedicated amateurs. Even after its emergence at the beginning of the nineteenth century as a fully independent branch of science, the appearance of a growing cadre of professional geologists stimulated rather than inhibited the activity of a still larger band of amateurs. In England especially, geology was the popular science *par excellence*. Clergymen, professional men, scholars, leisured gentlemen, and scientists from other branches of science were all attracted to its problems. They filled the geological societies and made important contributions to the literature of the maturing science. Kelvin was such an amateur.

Pre-eminently a physicist and engineer, Kelvin's interests in geology were never quite those of a geologist. Except for a brief period in 1849, he did virtually no field work and frequently ignored the purely geological data accumulated by others. He also ignored or was frankly oblivious to some of the most significant geological problems of his day. But though his geology was based upon physical rather than geological principles, the speculative, frequently inspired papers that he produced regularly for nearly sixty years touched upon the seminal problems of physical geology; and during much of that time, his ideas profoundly influenced the course of geological thought. In the case of the age of the earth in particular, they stimulated a conceptual revolution.

Preliminary Speculations

One of Kelvin's earliest scientific papers was an undergraduate essay at Glasgow entitled "On the Figure of the Earth." In several ways this essay illustrates in microcosm Kelvin's devotion to physical geology and two of the most important reasons for his success, tenacity and the unswerving application of physical principles. Significantly, the essay was based upon the works of mathematicians and astronomers rather than geologists, and it was good enough to win for him a university medal when he was sixteen. Perhaps even more significant, however, although the essay was written in 1840 and was one of the few products of Kelvin's pen that was never published, the manuscript itself bears several emendations: one written in 1844 while he was a student at Cambridge, one in 1866 while aboard the Atlantic cable-laying vessel *Great Eastern,* and one in 1907 only two months before his death.[1] A similar tenacity marked Kelvin's devotion to the problem of the age of the earth and no doubt accounts in part for his influence. Had he been content merely to present his arguments with the evidence available, they would probably have been swamped by the sheer volume of conflicting geological opinion. But instead, he reiterated each argument repeatedly, modifying them to reflect new data but always leaving the basic hypotheses intact. And for a while at least, each repetition seemed to strengthen their persuasive force.

By the time Kelvin returned from Cambridge to Glasgow in 1846 he had already begun to consider seriously the relationship between the earth's age and its internal temperature. His first published work, completed in 1841, had already demonstrated his mastery of Joseph Fourier's theory of heat conduction, the key mathematical tool in the most important of his subsequent arguments. Accordingly, his inaugural dissertation before the faculty at the University of Glasgow was on the "Age of the Earth and its Limitations as Determined from the Distribution and Movement of Heat within it." The topic for the dissertation was chosen by the Glasgow faculty, but according to his biographer Silvanus Thompson, it was "probably suggested" either by Kelvin or his father. It would be interesting to discover how Kelvin presented his arguments before the formulation of the first and second laws of thermodynamics, but no known copy of the dissertation exists. Again according to Thompson, Kelvin burned the manuscript the same day that it was delivered, and its exact contents are unknown.[2]

During the years immediately following his appointment at Glasgow, Kelvin played a major role in the formulation of the principles of thermodynamics. But though these principles were to provide the basis for all of his later work on the earth's age, he found little time during those busy years to address the subject directly. Nevertheless, the problem was never completely out of mind, and when he announced the discovery of the second law of thermodynamics in 1852, he turned immediately to its implications for the age of the earth.

Within a finite period of time past the earth must have

been, and within a finite period of time to come the
earth must again be, unfit for the habitation of man as
at present constituted, unless operations have been, or
are to be performed which are impossible under the
laws to which the known operations going on at the
present in the material world are subject.[3]

Before he could elaborate on the terrestrial problem, however, his
attention was diverted to the problem of solar energy. At the meeting
of the British Association in 1853, he heard James J. Waterston propose the hypothesis that the sun's heat was generated by the impact of
meteors falling into its surface.[4] He also learned of a similar proposal
made a few years earlier by his friend James P. Joule (1818-1889).[5] He
became an immediate, and almost solitary, convert and the following
spring presented his own, significantly modified, version of Waterston's hypothesis to the Royal Society of Edinburgh.[6]

Despite the work of Joule and Waterston it was still generally accepted that the sun's heat was in some way chemically generated. Yet a
fairly simple calculation showed that no known chemical process could
possibly have supplied the sun's enormous expenditure of energy even
for the relatively short period of historical time. Rejecting a chemical
explanation, therefore, Kelvin saw only two other conceivable explanations: either the sun is merely a hot body losing heat, or its heat must
constantly be replenished by the impact of meteors falling onto its
surface.[7] At this time (1854) he believed the sun to be a solid sphere,
and his study of Fourier's theories had shown that heat conduction
from the solar interior would be quite inadequate to maintain a high
surface temperature for any length of time. Thus the only alternative
seemed to be some form of Waterston's meteoric hypothesis.

It was well known that, in its orbit around the sun, the earth encoun-
ters many meteors which are apparently following circular or elliptical
paths, but as Kelvin was well aware their number is far too small for
their combined potential energy to account for the sun's heat. Consequently, he concluded that the meteors responsible for the solar energy
must already be inside the earth's orbit and spiraling slowly into the
body of the sun. He further suggested that these meteors were in fact
clearly visible from the earth as the Zodiacal Light, a phenomenon which
he described as nothing more than a "tornado of stones" stretching
from the sun almost to the earth's orbit. In his original hypothesis,
Kelvin assumed that upon contact with the resistance of the sun's
atmosphere, the meteors would slow down until their spiral gave way to

a sudden rush to the solar surface with an accompanying rapid production and radiation of heat. According to his calculations, the resulting increase in the sun's mass would not appreciably affect the earth's motion since the meteoric material involved was already within its orbit, and the change in the sun's size would be insensible to terrestrial measurement.

Kelvin's original communication emphasized the simplicity and adequacy of the meteoric theory. At that time, his concern was merely to defend the scientific validity of his hypothesis, not to develop its consequences. During the next few months, however, he discovered several interesting corollaries to his hypothesis, and in a series of addenda, expanded it and made it more sophisticated.

The problem lay in the requirement that most of the meteors responsible for the sun's heat be inside the earth's orbit. Obviously, such a conclusion meant that the meteors themselves must remain for long periods in slowly spiraling orbits, and this suggested to Kelvin the motion of a vortex—a concept just then enjoying a renewed interest among physicists seeking mechanical explanations for electromagnetic and gravitational phenomena. He therefore postulated that the meteors might be caught in an ethereal vortex about the sun, and from there proceeded to the conclusion that the sun's heat was not generated by the direct impact of solid meteors, since they would in fact be vaporized by the intense heat long before impact, "but by the friction in an atmosphere of evaporated meteors, drawn in and condensed by gravitation while brought to rest by the resistance of the sun's surface."[8] It was this solar vortex that suggested the possibility of calculating the sun's age.

Starting from the well-known fact that the sun rotates upon its axis, Kelvin estimated the rotational velocity of his hypothetical solar vortex and its effect upon the falling meteoric matter. From these estimates he determined the proportion of the sun's present mass which meteors from the vortex must have supplied in order to account for its present rotary motion, and calculated that about 32,000 years would be necessary for the influx of this material. Since he believed that the inertia of the falling meteors was the most probable cause of the sun's rotation, he concluded that the sun could only have been illuminated during the time required for it to obtain its present rotational velocity, or in other words, that it could have supplied the earth with heat and light for "not many times more or less than 32,000 years." On the same grounds, he estimated that its energy cannot be maintained for much longer than

300,000 years in the future.[9] The earth's history, he reasoned, must be similarly limited.

Except for this single brief calculation Kelvin had confined himself to a qualitative discussion of a broad general problem, and when he addressed the British Association later the same year (1854), he continued in much the same vein, including a summary of the earlier paper. One point in his address requires special attention, however, because it contains the most complete summary he was ever to give of the premises which were to dominate all of his subsequent speculations on the age of the earth. What, he asked his audience, is the fundamental *source* of all of the mechanical energy in the solar system? He then made his own reply:

> This is a question, to the answering of which mechanical reasoning may be legitimately applied: for we know that from age to age the potential energy of the mutual gravitation of those bodies is gradually expended, half in augmenting their motions, and half in generating heat; and we may trace this kind of action either backwards or forwards; backwards for a million of million years with as little presumption as forwards for a single day. If we trace them forwards, we find that the end of this world as a habitation for man, or for any living creature or plant at present existing in it, is *mechanically inevitable*; and if we trace them backwards according to the laws of matter and motion, certainly fulfilled in all the action of nature which we have been allowed to observe, we find that a time must have been when the earth, with no sun to illuminate it, the other bodies known as planets, and the countless smaller planetary masses at present seen as the zodiacal light must have been indefinitely remote from one another and from all other solids in space. All such conclusions are subject to limitations, as we do not know at what moment a creation of matter or energy may have given a beginning, beyond which mechanical speculations can not lead us. If in purely mechanical science we are ever liable to forget this limitation, we ought to be reminded of it by considering that purely mechanical reasoning shows us a time when the earth must have been tenantless; and teaches us that our bodies, as well

as all living plants and animals, and all fossil organic
remains, are organized forms of matter to which science can point no antecedent except the Will of a
Creator, a truth amply confirmed by the evidence of
geological history. But if duly impressed with this limitation to the certainty of all speculations regarding the
future and prehistorical periods of the past, we may
legitimately push them to endless futurity, and we can
be stopped by no barrier of past time, without ascertaining at some finite epoch a state of matter derivable
from no antecedent by natural laws. Although we can
conceive of such a state of all matter, or of the matter
within any limited space and have cases of it in the
arbitrary distributions of temperature, prescribed as
"initial" in the theory of the conduction of heat (...), yet
we have no indications whatever of natural instances
of it, and in the present state of science we may look for
mechanical antecedents to every natural state of matter which we either know or can conceive at any past
epoch however remote.[10]

He went even further in his conclusion. Besides asserting his belief
that all of the heat and motion of the universe originated in mechanical
actions of the same kind as those which now sustain them, he suggested
the possibility "that the potential energy of gravitation may be in reality
the ultimate created antecedent of all motion, heat and light in the
universe."[11] He came to accept this possibility as fact, and it became the
key to all of his later work on the ages of the sun and earth.

Kelvin's papers in 1854 attracted little attention, and his activities
between 1855 and 1860 left him little time to continue his speculations.
Nonetheless, he was considerably aroused in 1859 when, for a short
time, it looked as though his views on solar heat had been vindicated.
The French astronomer, Leverrier, announced the discovery of a
pronounced progression of the perihelion of Mercury which he attributed to the gravitational attraction of matter lying between Mercury's
orbit and the sun. Leverrier also suggested the possibility of a planet
lying between Mercury and the sun (the hypothetical *Vulcan*), but he
believed a collection of corpuscles or stones was more likely. Kelvin
immediately seized upon the discovery to support his hypothesis of
meteors spiraling slowly into the sun. His elation, however, was short-
lived. The following year further calculations showed that a mass of

meteors sufficient to supply the sun's heat would create a much greater perturbation of Mercury's orbit than the one Leverrier had discovered. Nonetheless, rather than immediately desert his hypothesis, which he knew might prove untenable, he called for a suspension of judgment until more knowledge was available.[12] But the hoped-for evidence was not to be forthcoming, and the problem of the perturbation of Mercury's orbit remained unsettled until the next century.

Kelvin did not wait so long, however. In December, 1860, an accident left him bedridden with a broken leg, and for several months he was forced to forego the active pursuits that had occupied so much of the previous decade. He thus found time to reassess carefully his solar hypothesis.[13] Almost immediately the desire to suspend judgment on the question of solar energy vanished as he replaced his original speculations with a more sophisticated form of the meteoric-gravitational theory. By summer, a preliminary statement on the new hypothesis was ready for the meeting of the British Association, but in Kelvin's absence its reading apparently went unnoticed.[14] It was consequently in the guise of a popular article, published in *Macmillans Magazine* in March, 1862, that the new solar hypothesis first attracted public attention, and it was in that form that scientists had to deal with it for the next fifty years.

Time and Thermodynamics

The article on solar heat in *Macmillans Magazine* contained the first of three arguments in which Kelvin believed he had shown conclusively that the earth's age was both limited by physical laws and calculable from physical evidence. The second, which also appeared in 1862, was a further elaboration upon his early interest in the earth's internal temperature. The third, which did not appear until six years later, took a different but related approach based on the dynamics of tidal friction. All three arguments shared certain fundamental assumptions: that all of the energy in the universe was gravitational in origin, that it is limited and calculable in quantity, and that it is constantly being dissipated according to the second law of thermodynamics. Within the limits of these assumptions, however, each argument approached the problem in a different way, and their quantitative results were never entirely consistent. They were nonetheless sufficiently in accord to convince Kelvin and his contemporaries of their general validity, and though he periodically modified the numerical results obtainable, he never abandoned the substance of either argument. They thus provided the basis for his enormous influence on geochronology.

Kelvin's return to the question of solar heat had been sparked in part by the realization that his 1854 hypothesis was untenable. Subsequent calculations had shown that the energy available from meteors presently falling into the sun's surface would be inadequate to supply the sun's heat even for the relatively brief period of historic time. In order to avoid giving up the gravitational explanation altogether, therefore, it was necessary to push it further back in time, or in other words, to accept the probability that the sun is at present an incadescent liquid body losing heat. The *Macmillans* article, entitled "On the Age of the Sun's Heat," was devoted to justifying this new assumption, explaining its causes, and elaborating on its effects. It was an admittedly speculative paper intended for a popular audience. Despite the numerical results given in its conclusion, it contained neither mathematical formulas nor calculations. It was nonetheless closely reasoned and persuasively written, and its impact was tremendous.

Kelvin made it clear at the outset that, however speculative, his intention was to seek a purely natural explanation for the sun's heat. He therefore began by eliminating impossible or highly improbable explanations including his own earlier hypothesis, and found himself faced with the proposition that since a finite sun can have neither an infinite store of primitive energy nor a constant chemical or mechanical source of supply, it cannot have existed as a constant source of radiant energy for an infinite period of past time.

> The sun must, therefore, either have been created as an active source of heat at some time of not immeasurable antiquity, by an over-ruling decree; or the heat which he has already radiated away, and that which he still possesses, must have been acquired by a natural process, following permanently established laws. Without pronouncing the former supposition to be essentially incredible, we may safely say that it is in the highest degree improbable, if we can show the latter to be not contradictory to known physical laws. And we do show this and more, by merely pointing to certain actions, going on before us at present, which, if sufficiently abundant at some past time, must have given the sun heat enough to account for all we know of his past radiation and present temperature.[15]

This brief statement embraces the essence of the philosophy of nature underlying the whole of Kelvin's approach to geological time.

The search for actions going on in the present which may have been sufficient in the past to account for the sun's primordial energy took Kelvin straight back to gravitation and the collision of meteors. Quoting from his own earlier paper, he asserted emphatically that, *"Meteoric action . . . is . . . not only proved to exist as a cause of solar heat, but it is the only one of all conceivable causes which we know to exist from independent evidence."*[16] He could conceive of no other possible explanation except chemical action and that was patently insufficient even for historical time. Indeed, he asserted that if the whole mass of the sun were undergoing the most energetic chemical process known, it could generate only about 3,000 years' supply of heat while the meteoric theory can easily account for 20,000,000 years' supply.

The meteoric theory, as Kelvin saw it, was a straightforward application of the principle of the conservation of energy on a cosmic scale. When two or more meteors, of whatever size, are separated in space they possess a calculable amount of potential energy due to their relative positions and to their mutual gravitational attraction. As the meteors are drawn together by this mutual attraction, their original potential energy is progressively converted into kinetic energy which upon collision is converted into heat. According to the principle of the conservation of energy, the heat thus generated must be exactly equivalent to the original potential energy of the system. Thus, if the original aggregate mass of the meteors is known and if they can be assumed to have been originally at rest relative to each other, the amount of heat generated by their collision can readily be calculated from Joule's mechanical equivalent of heat. Kelvin's new solar theory simply expanded this mechanical situation to include a large number of meteors of aggregate mass equal to the sun's and mutually attracting each other toward their center of mass. The relatively rapid collision of a mass of meteors equal to that of the sun would be extremely energetic and the resulting body would be an incredibly hot molten mass. Because he assumed the original meteors to have been at rest, all of the energy involved in this hypothesis would be gravitational, and because they were assumed to have been separated by large distances compared to their diameters, the energy generated was readily calculable.[17]

Kelvin's new version of the meteoric hypothesis was far from original. In the form he presented it, it was the inspiration of the German physicist Hermann von Helmholtz, but its origins actually go back to the nebular hypothesis of Immanual Kant and Pierre Laplace. Helmholtz and Kelvin had simply made the nebular hypothesis more specific and had injected into it the concept of energy. By coincidence, Helmholtz had initially proposed this form of the meteoric hypothesis in February, 1854, a month before Kelvin's paper "On the Mechanical Energies of the Solar System."[18] Kelvin was apparently ignorant of Helmholtz's work when he read his original paper, however, although he had read it before attending the British Association meeting a few month's later. In fact, by then he had calculated, as Helmholtz had not, that the theory could account for 20,000,000 times the heat annually radiated by the sun. Nonetheless, he rejected the theory at the time on the grounds that most of the heat generated would be dissipated immediately and could play no part in the sun's present activity.[19]

For several years Kelvin continued to reject Helmholtz's version of the meteoric theory, and even when forced by the discoveries of Leverrier to abandon his own hypothesis embraced the alternative with some reluctance. A letter written to his brother-in-law while preparing the original version of the new theory tells of his doubts about assuming shapeless detached stones to be the primitive form of matter. It was almost as if he were arguing with himself that he confessed that meteors do exist by the millions in space, and to consider them to be broken fragments of larger planets was mere hypothesis. Consistency required that since the discovery of the primitive state of matter is beyond man's power, if one is to find a probable beginning, he must start with the present condition of nature and reason back by analogy and strict dynamics. On those grounds he conceded, however reluctantly, that he must accept the "probable supposition that the sun has been built up by the falling together of smaller masses."[20] It is probably not insignificant, nonetheless, that Helmholtz visited Kelvin while the new hypothesis was being considered, but no record remains of their conversations.[21]

Before adopting Helmholtz's theory, however, Kelvin still had to answer his own objections. For example, it remained necessary to allow for an appreciable energy loss during the process of collision and consolidation, and his estimates indicated that even in the case of the most rapid conglomeration of the meteors possible, about half of the original potential energy of the system would be dissipated almost immediately. On the other hand, it seemed highly probable that the sun's density increases greatly with depth, a condition that would allow for far more heat being generated during its formation than the calculation from uniform density would indicate. Bearing both these possibilities in mind, he placed the lower limit of the original supply of solar energy at 10,000,000 times the present annual heat loss while admitting that the upper limit might be five or even ten times greater. There is little doubt that Kelvin believed that these calculations had definitely established limits for the possible age of the sun, but because of the admitted uncertainty of his data, his final statements are cautious and, in retrospect, prophetic:

> It seems, therefore, on the whole most probable that the sun has not illuminated the earth for 100,000,000 years, and almost certain that he has not done so for 500,000,000 years. As for the future, we may say, with equal certainty, that inhabitants of the earth cannot continue to enjoy the light and heat essential to their life, for many million years longer, unless sources now unknown to us are prepared in the great storehouse of creation.[22]

Considering the limited knowledge of solar physics at the time, the theory outlined above gave the best available explanation for the *origin* of the sun. It was a logical extension of the widely accepted nebular hypothesis of Laplace and Kant and of the principles of thermodynamics which had been developed during the preceding decade. The question of the sun's *present activity*, on the other hand, was not quite so simple. While Kelvin had rejected the probability of meteors currently supplying all of the sun's energy loss, they must, if his theory were correct, still be supplying it with some energy. There was consequently no way of really ascertaining definitely whether the sun was really cooling at all. If it were cooling, as Kelvin believed, then other effects also contributed to its heat supply, chiefly the amount of potential energy converted to heat by its thermal contraction. The calculation of any heat generated in this manner, however, depended upon knowledge of such factors as the sun's density, its specific heat, and the amount of its contraction, all of which were unknown and beyond immediate determination. In other words, the already wide limits for the possible age of the sun had to be further broadened to allow for the uncertainity about present solar activity.

Kelvin was well aware of the inexactness of his quantitave estimates, and contrary to the dogmatic assertions of some of his later work, he was quite cautious in the way he put forth his conclusions. He had been drawn into the problem by his interest in the implications of the second law of thermodynamics for theories about the continued activity of the sun, and in spite of the uncertainties in his calculations, he believed that they presented adequate refutation for any theories of the earth's age that required immeasurably vast times. His conclusions placed him in opposition to the prevailing trend of uniformitarian geology at the time, and thus in his early papers he was less concerned with the exact age of the sun and the earth, than with rescuing geological theory from the violation of established physical principles. The inexactness of his calculations was consequently unimportant so long as they established the necessity for a limit upon geological time and the impossibility of uniformitarianism's demand for limitless ages.

In this regard, it is difficult to avoid the suspicion that the publication of Charles Darwin's *On the Origin of Species* played a part in stimulating Kelvin's renewed interest in the earth's age. Although his interest in the problems involved clearly predates Darwin's famous work, he had previously ignored the uniformitarians' demands for time. In 1862, however, he confronted Darwin directly:

> What then are we to think of such geological esti-
> mates as 300,000,000 years for the "denudation of the
> Weald?" Whether is it more probable that the physical
> conditions of the sun's matter differ 1,000 times more
> than dynamics compel us to suppose they differ from
> those of matter in our laboratories; or that a stormy
> sea, with possibly channel tides of extreme violence,
> should encroach on a chalk cliff 1,000 times more
> rapidly than Mr. Darwin's estimate of one inch per
> century?[23]

In Kelvin's view the choice was obvious. Darwin was wrong! Nonetheless, he apparently appealed to John Phillips (1800-1874), the professor of geology at Oxford, for a geological evaluation of Darwin's remarks on the age of the Weald. Phillips, who had been an early critic of Darwin's geology, replied with characteristic bluntness. "Darwin's computations are something absurd," he proclaimed. Indeed, his own calculations indicated that the total age of the earth's entire stratified crust was only 96 million years.[24] Needless to say, Kelvin found this figure remarkably congenial with his own results.

Why Kelvin chose Darwin rather than Lyell for his initial assault on uniformitarianism raises several interesting speculations. Certainly he knew the outlines of Lyell's position. Given his own interest in the subject, he could hardly have failed to have followed the debate between Lyell and William Hopkins during the early fifties,[25] particularly since Hopkins had been his mathematics coach at Cambridge only a few years earlier. Nonetheless, he seems not to have fully recognized the implications of Lyell's doctrine for geochronology. He probably had never read Lyell systematically,[26] and even if he had, Lyell always spoke rather vaguely of "indefinite" or "limitless" expanses of time, never in concrete numerical terms. For Kelvin a million years was clearly vast enough to be inconceivable or "indefinite" on a human scale. In fact, he casually mentioned in 1854 that two million years should be ample for the speculations of geologists.[27] Darwin, on the other hand, had boldly referred to the age of the Weald in years, and had provided a concrete calculation that Kelvin believed could be directly refuted by physical reasoning.[28]

Certainly, too, there can be little doubt that natural selection itself provided an additional spur to Kelvin's activity. Although he made it clear in 1871 that he was not opposed to evolution as such, he rejected natural selection because it left no place for the operation of design or divine order in the evolution of life.[29] And for Kelvin, design was as much a principle of nature as were the laws of thermodynamics. Indeed, it was the belief in design that justified the formulation of universal scientific laws, that assured the relationships of cause and effect, that, in short, made science possible. Whatever the shortcomings of his treatment of geological time, Lyell shared Kelvin's belief in the reality of design in nature, whereas Darwin seemed to threaten it and thus drew to himself the brunt of the physicist's attack.

Not content with a single line of attack, Kelvin's second argument followed hard on the heels of the first. As its title suggests, "On the Secular Cooling of the Earth" was an attempt to treat the earth as a cooling body losing the heat generated in the primordial collision of the matter from which it was formed.[30] As in the case of the sun, Kelvin's approach was physical, and there were obvious parallels between the basic assumptions involved in the two cases. But the difference between the two problems was significant. Whereas in the case of the sun Kelvin had been forced to deal in cosmic terms with the total energy available and the total time since its formation, in the case of the earth he could confine himself to geological time which he defined as the time since the formation of a solid crust. Perhaps even more important, the parameters used to determine the sun's age had all been mere estimates or at best the results of indirect measurement, whereas several of those needed for the terrestrial calculation could be measured directly.

It seems probable that Kelvin was first attracted to the problem of the earth's internal temperature through his early interest in the work of Fourier.[31] Certainly by the time he delivered his inaugural dissertation at Glasgow in 1846, he was already convinced that the distribution of heat within the earth contained the key to determining its age. In 1849 he even began to gather data on the subject by assisting James David Forbes of Edinburgh in the measurement of underground temperatures and the thermal properties of rocks. As in the case of the sun's heat, however, other activities interfered, and though he continued to develop his ideas, it was not until 1861 that he found time to put them all together.[32]

The fundamental idea that the earth had originally formed as a hot molten ball and had cooled gradually to its present condition was of course far from new. Observations in countless mines and tunnels had shown that the earth's temperature increases with depth and consequently had suggested that it must be losing heat by conduction through its outer surface.[33] As early as the seventeenth century, in fact, both Descartes and Leibniz had discussed the cooling of the earth from a molten state, and in the eighteenth, Buffon had even made a crude effort to determine its age from its rate of cooling.[34] Kelvin's accomplishment was to impart an air of precision to the problem.

Since the terrestrial problem dealt only with geological time, it was not necessary to account for the origin of the earth so long as it could be shown that an originally molten condition was both physically possible and reasonably probable. Kelvin contented himself with suggesting that either the impact of large numbers of meteors upon an initially cold nucleus or the collision of two bodies of approximately equal masses could have produced the necessary conditions. In either case the ultimate source of the earth's heat would be gravitation. A much more fundamental question was how the earth's solid crust had formed, for upon that depended the critical assumptions needed to determine its rate of cooling. Kelvin consequently examined his alternatives carefully.

Looking first at the possibility that molten rock might expand upon solidifying as water does upon turning into ice, he reasoned that the earth's solid outer crust would form over a still molten interior in the way ice forms on a lake. The resulting fluid core model of the earth was essentially the same as the one long accepted by many geologists because of its flexibility in explaining seismic phenomena. The limited evidence available, however, indicated that liquid rock actually contracts rather than expands upon solidifying. And even more significant from Kelvin's point of view, William Hopkins had for years opposed the hypothesis of a fluid terrestrial core on the grounds that it was totally inconsistent with the physical and astronomical evidence for the earth's rigidity.[35] Kelvin's own calculations, moreover, placed him firmly in Hopkins' camp. Assuming therefore that solid rock must be more dense than liquid rock a second possibility was that slabs of rock would solidify in the cooler temperatures at the earth's surface and then sink towards its center, building up in the process a comb-like structure which would eventually support the outer crust. At various times, Kelvin gave some credibility to this model since it would account for the earth's structural rigidity while allowing for large reservoirs of molten lava in its interior. But on balance, he believed a third alternative to be the most probable. Assuming that molten rock contracts upon cooling as well as upon solidifying, he suggested that rock cooling near the earth's surface would sink before solidifying, thus creating convection currents which would maintain thermal equilibrium throughout the whole globe until solidification began. The earth would then solidify from the center outwards, and at the time the crust formed, would consist of a solid sphere at a uniform temperature throughout.

Adopting the third alternative, Kelvin next faced the task of marshaling the limited data available to him. His proposed model provided an ideal case for Fourier's mathematics, but meaningful results required quantitative values for the earth's internal temperature, the thermal gradient at its surface, and the thermal conductivity of its constituent rocks. Of these, the second seemed to present the least difficulty. Enough measurements had been made by J.D. Forbes and others to show that the thermal gradient varied with locality, but a temperature increase of 1°F in 50 feet was generally accepted as a probable mean. As for thermal conductivity, Kelvin had himself measured the conductivities of numerous surface rocks and was convinced that they provided an adequate average for the earth as a whole. Only the earth's deep internal temperature defied direct measurement, and there the assumed method of solidification became significant. If, as Kelvin assumed, the earth solidified at a uniform temperature throughout, that temperature should have been very near the normal fusion temperature of surface rock. It should also be the present temperature of the earth's uncooled inner core. Again citing the authority of general acceptance, Kelvin adopted a probable value for the fusion temperature of rock as his third parameter.

Kelvin was well aware of the drawbacks involved in arguing from such evidence. He knew that his data were themselves dependent upon the assumptions made in setting up his calculations, that they were at times little more than refined guesses, and that they were subject to numerous corrections. At one point he even ruefully admitted that physics was still "very ignorant as to the effects of high temperatures in altering the conductivities and specific heats of rocks, and as to their latent heat of fusion."[36] Nonetheless, he was fully convinced that the data were based upon the best measurements available and that they were sufficiently exact to make meaningful calculations possible; and he consequently devoted the bulk of his paper to defending them and to justifying his assumptions on the basis of known physical principles. When he came to the final calculation, therefore, he found that the data and assumptions as presented would yield an age of the earth's crust of 98 million years, but he felt it prudent to extend the acceptable limits to between 20 million and 400 million years to allow for the necessary uncertainties involved. In Kelvin's terms, this was an extremely cautious result, and his careful defence of the principles on which it was based made it convincing indeed.

"On the Secular Cooling of the Earth" was concerned with much more than the mere calculation of geological time, however. As Kelvin candidly admitted, it was written in order to show that geologists, especially uniformitarian geologists, had neglected the principles of thermodynamics in their speculations. His argument hinged on the assumption that the heat which has obviously been lost by the earth has resulted in its actual cooling, and consequently that no theory can possibly be valid that postulates a "steady-state" earth or one in which essentially identical geological cycles are constantly repeated. If, on the other hand, some internal or external source of energy could be shown to provide all or some portion of the energy lost by the earth, the whole argument would itself be invalid. But Kelvin could conceive of no such possibility and devoted several pages to showing that no other source was remotely feasible. He took particular issue with Lyell on this point, singling out for special criticism the suggestion that a self-sustaining cyclic exchange of chemical, electrolytic, and heat energy might occur within the earth. This proposal, which amounted to postulating a perpetual motion machine, was directly contrary to the second law of thermodynamics and thus demonstrated, in Kelvin's opinion, the extent to which geologists were guilty of ignoring physical principles.[37]

Going still further into the fundamentals of the uniformitarian-catastrophist controversy itself, Kelvin admitted that he believed that those theories requiring great geological catastrophes were *probably* false, but in much stronger terms he denied the *possibility* that geological and meteorological activity could remain constant for even so long as a million years. The sun and the earth are both losing energy, he argued, and there is no conceivable way in which it is being replaced. The sun must therefore have been hotter in the past than at present, and the increased solar radiation received by the earth must have resulted in greatly accelerated rates of evaporation, precipitation, and land erosion. By the same logic, it follows that a higher temperature of the earth in the past must have generated greater levels of volcanic and plutonic activity than presently exist. Kelvin did not feel that the acceptance of these conclusions constituted a return to the doctrine of great geological catastrophes. Although he did not explicitly state his view, it is clear that he interpreted catastrophism to mean the interference of supernatural forces in geological processes. Thus he did not regard greatly accelerated geological actions in the past as catastrophes, no matter how violent their effects, so long as they resulted from natural causes. In this sense, Kelvin believed himself to be the defender of a middle ground. Although he attacked only uniformitarianism directly, his broader aim was to insure that the results of geological speculation be made physically and philosophically sound. This was a purely scientific goal, and any return to supernatural catastrophes would be as repugnant as a continuation of radical uniformity.

By the time Kelvin's third argument appeared, his earlier works were well known and had already begun to affect the calculations of a few geologists. Kelvin was not satisfied, however. Uniformitarianism still dominated British geology, and although the meaning of the terms uniformity and uniformitarianism were being questioned and even modified by some geologists, others, notably Lyell, still refused to curtail their demands for an indefinite expanse of time. It is not surprising, therefore, that when Kelvin addressed the Geological Society of Glasgow early in 1868, his topic was "On Geological Time."[38] The address was a direct frontal assault on uniformitarianism, but for reasons not made clear, he chose to direct his remarks at the work of John Playfair, rather than at the more recent work of Lyell. Continually, he raised points made by Playfair only to refute them. He reiterated his earlier arguments for limiting the age of the earth, but they were mere afterthoughts to enforce his central theme. The bulk of the address was devoted to developing still a third method for determining the limits of geological time, a method based upon calculating the effect of the tides in retarding the diurnal rotation of the earth.

As he had in the earlier arguments, Kelvin took his starting point from established phenomena and familiar dynamic principles. In this case he drew upon the well known fact that the tides lag considerably behind their theoretically ideal positions. It had been one of the earliest conclusions drawn from the theory of universal gravitation that the tides resulted from the gravitational attraction of the sun and moon acting upon the waters of the earth. Ideally, high tide should always occur at the points on earth closest to and diametrically opposite from the moon with lesser tides at corresponding positions with respect to the sun. But because of the friction of the water, including both the internal friction due to the water's viscosity and the surface drag along the shores and sea bottom, the true tides lag behind these ideal positions. The principle of conservation of energy requires that the energy lost to heat by this continual frictional drag be accounted for, and it was Kelvin's suggestion that the loss must be at the expense of the energy of the earth's axial rotation. In other words, the tides must slow down the earth's axial rotation in the same way a frictional brake slows down the turning of a wheel.

Kelvin claimed no originality for this explanation. It had been suggested to him first by his brother James, and he later discovered that it had been previously discussed by Kant, Mayer, and Helmholtz.[39] He had also no doubt been influenced by the astronomer John Couch Adams' discovery of a change in the relative angular velocities of the earth and moon, the magnitude of which he had been asked to help calculate.[40] His originality lay in the way in which he related what was essentially a problem in physical astronomy to the question of the age of the earth.

In order to make this critical transition Kelvin had to go back to the postulates of his earlier arguments, namely to the assumption that the earth was at one time a molten ball but is now many times more rigid than steel. He believed it to be impossible that the present rigid earth could be deformed by the centrifugal force of its rotation, yet since the eighteenth century it had been known that the earth was indeed flattened at the poles and slightly expanded at the equator. The solution seemed quite obvious. At the time of its solidification, the earth must have taken on the shape of the spheroid of revolution corresponding to its angular velocity. But since the earth's shape does not differ greatly from the spheroid corresponding to its present velocity, Kelvin concluded that it could not have solidified very long in the past. Furthermore, he suggested that if one could determine how much the earth's rotation decelerated annually, he could extrapolate back and find the time at which the earth must have first solidified. Kelvin was not at the time prepared to calculate the earth's age on these grounds with any degree of precision, but he felt that he could place broad upper limits which would be enough to invalidate the conclusions of uniformitarian geology.

Qualitatively, Kelvin's method was straightforward and relatively simple. Quantitatively, it was made complex by the uncertainties in the data and the many corrections necessary to account for the perturbing influences inherent in any problem in gravitational astronomy. As Kelvin himself admitted:

> It is impossible, with the imperfect data we possess as to the tides, to calculate how much their effect in diminishing the earth's rotation really is . . . but it can be shown that the tidal retardation of the earth's rotation must be something very sensible.[41]

Part of the problem was that the diminution of the earth's rotation was *not* "very sensible" when measured by the only astronomical clock convenient for its determination, the revolutions of the moon. Laplace, for example, had used ancient Babylonian records of lunar eclipses to show that the relative velocities of the earth and moon had not changed appreciably in several thousand years. Kelvin was able to explain away the anomaly, but only at the expense of revealing why accurate calculations were impossible. According to Newton's third law, as the moon exerts a retarding force on the earth's rotation, the earth must at the same time exert a reaction force increasing the moon's *moment of motion*. Instead of increasing the moon's angular velocity, as one might at first assume, however, the increased moment of motion results in an increase in the moon's distance from the earth and a consequent decrease in its apparent angular velocity. Thus, the earth and the moon tend to slow down together but at slightly different rates.

Other factors would also tend to limit the effectiveness of the earth as an accurate chronometer, and in an earlier paper Kelvin had considered two of these, both of which were related to his theory of the earth's cooling. First, meteoric matter arriving at the earth's surface, if it arrives at an appreciable rate, should increase the earth's moment of inertia and decrease its velocity of rotation. His calculations showed, in fact, that a layer of meteoric dust increasing at a rate of 1/20th of a foot per century would retard terrestrial rotation by about eleven seconds per century, or roughly the amount by which Adams' calculations showed the moon to be gaining on the earth. On the other hand, if the earth were shrinking upon cooling, as Kelvin thought it must, its moment of inertia must decrease and thus its velocity of rotation would increase. The net effect of these and other possible perturbations remained unknown.[42]

Considering the possibilities for uncertainty and error, Kelvin's reluctance to try to impose definite limits on the age of the earth from tidal retardation is quite understandable. Nonetheless, the argument does illustrate his belief that if he could place any limit on the age of the earth, he had effectively refuted uniformitarianism. Here as in his earlier arguments, he did not seek to align himself with either group in the old catastrophist-uniformitarian debate, but sought instead to bring geological speculation into line with what he considered to be the irrefutable laws of natural philosophy. He was content to assert that if the earth consolidated one hundred million years ago its shape may well have been about what it is now, but if it had solidified ten thousand million years ago—which he thought would not satisfy some geologists—it must have been rotating twice as fast as at present and could not possibly have assumed its present shape. He felt that his conclusions really need go no farther. His earlier arguments had produced what he considered to be reasonable quantitative limits to the earth's age, and the argument from tidal retardation supported those conclusions. The new argument had not refined the precision of his calculations, but he believed that even the broadest of his chronological limits refuted uniformitarianism. For the time being, it was enough.

The Later Arguments

The three lines of argument discussed above exerted a lasting influence upon the development of geological theory. Their impact was intensified, moreover, by the fact that Kelvin returned to them throughout his long and productive life. Generally his later pronouncements were mere reiterations of the arguments formulated during the sixties stated in more positive terms to reflect the increase in the apparently confirmatory data at his disposal. With each new pronouncement he lowered the permissible upper limit of the earth's age as he reduced the allowance for error that he believed his calculations required. This change in numerical results proved to be significant.

Geologists, who were willing to submit to his mathematics and accept 100 million years as the earth's age, rebelled against these more restrictive later estimates.

The argument concerning tidal friction was the one Kelvin altered least. During the seventies particularly, he returned several times to the questions of the earth's solidity and rigidity, but gave little attention to the chronological implications of his results other than to reaffirm his conviction that they refuted uniformitarian geology. His original hypothesis required the development of an adequate theory of tides, but instead of developing a theory of his own, he encouraged the work of his younger colleague, George H. Darwin. From the beginning, the argument from tidal retardation had produced what Kelvin considered to be the least exact quantitative estimate of the earth's age, and it continued to be unreceptive to quantitative modification. In fact, Darwin's work seemed to make the chronological limits even less precise than before. Nonetheless, Kelvin remained convinced that tidal retardation would produce results which were consistent with those from his other arguments, although he made no attempt to refine his original statement.[43]

He did not similarly abandon the problem of the sun's heat, however. Throughout the sixties and seventies he referred to it again and again, taking his interest well beyond geochronology. The sun's heat, he believed, was the ultimate source of all the energy on earth except for tides and volcanic activity. He consequently made frequent allusions to the source and limitations of the sun's heat, especially in popular lectures and addresses, and his conclusions were invariably the same: the sun's energy must be limited by the original gravitational potential energy of its constituent parts, and its ability to make the earth habitable must be similarly limited.[44]

During the eighties, however, Kelvin's hypothesis encountered a serious obstacle. His original theory of a cooling sun implied a higher solar temperature in the past as well as a greater store of energy. There should, therefore, have been a corresponding increase in the intensity of the effects of solar radiation at the earth's surface. Climates should have been warmer and meteorological events—storms, winds, and rains—more violent. Yet, geological evidence showed no change in the magnitude of these events since the earliest Cambrian strata were laid down. As for climates, they had indeed been warmer in the past, but the Ice Ages bore witness to the fact that they had also been colder.

Kelvin sought to unravel the dilemma, and in 1886 he spent a few days with George Darwin discussing the problem. His failure is apparent in the card he sent Darwin shortly afterward:

> Chemical energy is infinitely too meagre in proportion to gravitational energy in respect to matter falling into the sun, to be almost worth thinking of. Dissociational ideas are widely nugatory in respect to such considerations. Hot primitive nebula is a wildly improbable hypothesis—so it now seems to me at all events, and can help us in nothing.[45]

He announced his solution the following year. In a lecture at the Royal Institution, he proposed an intricate mechanical analogy to illustrate how a slight cooling of the sun might generate almost as much heat as it lost through radiation. The model, which was an unacknowledged elaboration on a suggestion made in Helmholtz's paper of 1854, explained how the thermal contraction of the sun on cooling would convert some of the potential energy of its parts into heat. The sun was therefore not the simple cooling body of Kelvin's earlier hypothesis, but a vast storehouse of energy which is slowly being released while its temperature remains nearly constant.[46]

It seems evident here that in spite of the apparent dogmatism of some of his pronouncements, Kelvin had not closed his mind completely to the arguments and discoveries of the geologists. Nor had he fallen into the uncritical repetition of conclusions arrived at years before. Clearly on this point he had attempted to reinterpret his earlier hypothesis; but just as clearly, the reinterpretation had failed to sway his conviction about the age of the sun. The sun's total energy remained limited by the total primordial energy of its parts, and the limitation of its age remained unchanged. "In the circumstances," he concluded, " . . . it would, I think, be exceedingly rash to assume as probable anything more than twenty million years of the sun's light in the past history of the earth, or to reckon on more than five or six million years of sunlight to come."[47] Instead of extending the sun's duration, this reassessment of the problem had reinforced Kelvin's earlier conclusions about geological time. Still, it seems significant that in writing about the sun's heat in 1892, he repeated the exact words he had used forty years before:

> Within a finite period of time past the earth must have been, and within a finite period of time to come must again be, unfit for the habitation of man as at present constituted, *unless operations have been and are to*

be performed which are impossible under the laws governing the known operations going on at present in the material world.[48]

The theory of the secular cooling of the earth differed from the other arguments in that it was the one most amenable to quantitative revision. The argument itself seemed unshakable, but due in part to the efforts of the British Association's Committee on Underground Temperatures, information about the earth's thermal condition increased markedly. Kelvin, a perennial member of the committee, became ever more convinced of the validity of his arguments and of the accuracy of his calculations. With almost harmonic regularity he returned to the related questions of the earth's age and its internal temperature, and each time geology felt the pinch of a more restrictive chronology. In 1862 the uncertainties of his data led him to admit at least the possibility of a 400 million year old earth. By 1868, he was convinced that sufficient evidence existed to justify limiting the assumed duration of life on earth, if not the earth's total age, to no more than 100 million years. In 1876 he was willing to accept an upper limit of only 50 million years for the earth's age, and in 1881 a limit of 20 to 50 million years. Finally, in 1897 he declared that the earth's age was nearer 20 million than 40 million years, and embraced Clarance King's estimate of 24 million as the best available.[49] For a time geologists, convinced or cowed by Kelvin's mathematics, yielded to his results. But as the estimates grew more restrictive they grew more restive. In 1897, for the first time, Kelvin's results from his theory of terrestrial cooling were nearly in agreement with those from his theory of solar heat, but as he was well aware, neither was in close agreement with the hypotheses of the geologists.

Kelvin's willingness frequently to alter his numerical estimates of the earth's age, even to admit the possibility of a considerable expansion or contraction of his results, may seem inconsistent with his dogmatic refusal to alter his original propositions, but the inconsistency becomes more apparent than real once his basic position is recognized. Although his quantitative results greatly influenced geological thought and profoundly troubled some geologists, they were of secondary importance to their author. Of primary importance, in his opinion, were the inviolability of basic physical laws and the validity of geological inferences made from them. As long as the estimates of the earth's age drawn from his arguments were finite and of sufficiently short duration to refute uniformitarianism, he left them relatively unquestioned.

He was thus unperturbed when it became necessary to modify one of his quantitative chronological estimates or when the results from his various arguments differed greatly.

Two instances vividly illustrate Kelvin's attitude toward his numerical estimates of the earth's age. In 1895, John Perry, one of his former students, challenged some of the basic assumptions of the secular cooling argument. Kelvin readily conceded that his theory may have made too few allowances for unknown factors, and that based upon terrestrial cooling alone, the earth's age might be 4,000 million rather than 400 million years. But, he went on, the theory of solar heat immediately contradicted this conclusion and limited geological time to some small multiple of 20 million years.[50] Obviously, Kelvin was confident that what he had to yield on one side he would regain on the other. His continued interest in the problems of the earth's rigidity, however, led to a less equivocal compromise. In 1876, in response to a growing conviction among geologists that the earth's interior must be viscous, Kelvin reexamined his calculations on terrestrial rigidity. His results left him convinced that the earth's interior must be both solid and rigid, but they also indicated that it might be necessary to lower the accepted estimate of the earth's internal temperature, and this, in turn, would mean reducing his estimate of the earth's age. It is significant that Kelvin was quite willing to sacrifice his estimate of the earth's age—accepting 50 million years as a maximum—rather than abandon his belief in its rigidity. Physical reasoning had convinced him that the astronomical behavior of a partially fluid earth would be inconsistent with the laws of dynamics as he understood them, and he was unwilling to make such a concession to geological speculation.[51] The problem of the age of the earth was important to him because it demonstrated the strength of physical laws and physical reasoning. He was consequently content to let geology make what compromises it must.

Kelvin so successfully shifted the burden of proof to those who opposed his geochronological limits, that he seldom felt his arguments threatened in spite of frequent critical attacks. His arguments were based on strict dynamic reasoning, they followed the known laws of physics, and they utilized the most complete data available. Only if an argument proved wrong on its own terms would he abandon it, and that occured only twice. He adopted Helmholtz's solar theory and abandoned his own early meteoric hypothesis because the latter proved inconsistent with Leverrier's astronomical discoveries; and forty years later when he recognized the implications of radioactive heating, he reluctantly abandoned his purely mechanical explanations of solar and terrestrial heat.[52] During the years between those two instances, he occasionally modified his result to take advantage of new data, but his basic position remained unchanged. There were critics, of course, who pointed to the uncertainties in his data and warned against his mathematics giving an air of precision to uncertain results. But as geological evidence seemed increasingly to converge upon chronological limits near those of physics, the criticism rapidly declined. Throughout the whole period, his physics and mathematics seemed unassailable. It is little wonder, therefore, that his statements grew more emphatic, that his conclusions carried such conviction, and that his influence was so pronounced.

The Age of the Earth and the Laws of Nature

Kelvin's influence upon nineteenth century ideas about the age of the earth can hardly be exaggerated. As we shall see, he almost single-handedly forced geologists to revise the meanings of the terms "uniformity" and "uniformitarianism," and for nearly fifty years imposed his limitations upon virtually every attempt at establishing an exact geochronology. The reasons for his success were manifold, but a few factors in particular seem to account in great measure for his influence. In the first place, his fame and personal prestige lent weight to his hypotheses and assured them of an attentive hearing. Secondly, the arguments themselves were irrefutable in terms of the physics of the nineteenth century. But probably more important than either of these, his arguments were based upon a consistent and fairly complete cosmological system that was remarkably congenial to Victorian scientific thought. Although it was neither complex nor profound, this cosmology depended upon three fundamental postulates—the existence of design and order in the universe, the validity of the nebular hypothesis or some modification of it as the only probable cosmogonal theory, and the reality and universal applicability of natural laws—which were all widely accepted by Victorian scientists. Kelvin's views thus demanded attention even from those whom he attacked for disregarding certain natural laws. After all, the discovered laws of nature had grown so complex that they required interpretation by an authority, and he was the authority.

Modern historians of science tend to relegate Kelvin to a secondary position among the luminaries of Victorian science. They see him as a man who spread his talents too thinly and who failed to develop his ideas into a lasting synthesis. His contemporaries, on the other hand, admired the breadth of his knowledge and recognized its depth as well. To them, he was a man whose scientific ingenuity, intuition, and perception had no peer except Newton himself.[53] Kelvin was only thirty-eight when he published "On the Secular Cooling of the Earth," but for more than a decade he had been regarded as one of Europe's foremost scientists. Moreover, his fame and prestige increased steadily throughout his life until his authority came to be almost unquestioned in physics, mathematics, and engineering. He was particularly at home in public gatherings where his quick grasp of problems and intuitive insights could have full sway, and where his influence could have immediate and direct effect. As J. J. Thomson recalled:

> His personality was as remarkable as his scientific achievements; his genius and enthusiasm dominated any scientific discussion at which he was present. He was, I think, at his best at the meeting of Section A (Mathematics and Physics) of the British Association. . . .He made the meeting swing from the start to finish, stimulating and encouraging, as no one else did, the younger men who crowded to hear him. Never had science a more enthusiastic, stimulating, or indefatigable leader.[54]

But Kelvin was more than the acknowledged leader of British physical science. With his interest and success in both the practical and the theoretical aspects of science, he was the archetypal example of the successful Victorian scientist. It is little wonder, therefore, that his geological arguments, based as they were upon mathematical and physical reasoning, should exert so profound an influence upon his contemporaries. Kelvin's eminent position assured an audience for his geological speculations, but even more, it demanded that they be considered in any other attempt to set up a geological time scale.

Once he had found his audience, the undoubted accuracy of Kelvin's mathematics and the widespread acceptance of his presuppositions helped to create an air of invincibility about his arguments which greatly enhanced their authority. If his presuppositions were granted, his conclusions were both logical and legitimate.[55] Admittedly the presuppositions themselves may have been beyond proof, mere educated guesses, but before the discovery of radioactivity they were also beyond disproof. Moreover, they were based upon the most widely held hypotheses and the best empirical data available at the time. Thus as late as 1895 when John Perry finally challenged Kelvin's results, he could not prove that the underlying assumptions were false; he could only show that they were not necessarily true. And as Perry pointed out, his mathematics were even more unassailable.[56] His calculations were both rigorous and elegant, and they gave his conclusions a convincing aura of precision.

The illusion of exactness thus imparted by Kelvin's calculations seems to have particularly impressed geologists, biologists, and laymen who were unfamiliar with mathematics. But even physicists who were aware of the imperfections in the available data and the consequent inexactness of Kelvin's conclusions, still accepted the truth of his reasoning and the validity of his fundamental results. To take only one example, in his review of *On the Origin of Species,* Fleeming Jenkin stressed the fact that Kelvin's results were mere approximations, and yet confidently used them to oppose Darwin's theory.[57] It seems clear, however, that Kelvin's mathematics, his unproved but irrefutable assumptions, and even his immense prestige could not in themselves account for his influence on geology. In attacking the views of Charles Lyell, after all, he had attacked a man whose reputation in geology was at least equal to his own in physics. Part of the reason for his success, therefore, must lie in the receptivity of his audience, and in particular, in the fact that the scientists of the second half of the century were seeking a greater degree of quantitative precision that uniformitarianism's vague conception of indefinite time could provide. Kelvin's hypotheses, with the apparent internal consistency of their underlying assmptions, could provide the desired aura of exactitude while remaining consistent with the broader framework of Victorian scientific thought.

Of the three fundamental postulates which underlay Kelvin's approach to the age of the earth, the validity of two, the existence of design in the universe and the nebular hypothesis, was generally acknowledged during the late nineteenth century. The third postulate, the universal applicability of natural laws—whether these laws were interpreted merely as operational inductive generalizations or as the work of a Divine Artificer—was also almost universally accepted. It was in his development of a specific aspect of the latter postulate, however, and especially in his emphasis that the laws of thermodynamics must apply to any physical system including the universe as a whole, that Kelvin made his unique contribution to the problem of geochronology. He was singularly qualified for this task since he had been greatly

responsible for the establishment of the laws of thermodynamics and
for furthering their acceptance.

Like virtually every other scientist of his time, Kelvin had read
Paley's *Natural Theology*. Less typically, however, as late as 1903 he
remarked that upon re-reading Paley's book he found that he still
agreed with its arguments. There can be little question that Paley's
work and the prevalent belief in the evidence of design in the universe
made a lasting impression. Kelvin's papers on the age of the earth make
frequent reference to teleological arguments, and until the end of his
life he regarded the regularity and order of the universe as evidence of
the work of a guiding intelligence. Nonetheless, although his teleological views are essential to the understanding of his prolonged interest in
the question of terrestrial chronology, they must be approached with
caution. They should not be confused with the more specifically religious views which played such important roles in the attacks upon
Darwinian evolution and in the earlier uniformitarian-catastrophist
controversy. Kelvin, with justification, would have disclaimed being
agnostic, but on the other hand neither was he particularly theistic.
Silvanus Thompson's account of his religious life presents a picture of a
man who was personally, but not conventionally, religious; a man who
would argue for the opening of museums, galleries, and libraries on
Sunday, but who would dock his yacht on Sunday morning so that the
crew (but not necessarily himself) could attend church. He was equally
at home at Presbyterian services in Glasgow, Episcopal services in
London, and Free Church services at Largs. Kelvin's religious views
were no doubt important, but they were neither fundamentalist nor
tied to Biblical literalism, and they were not, at least in any strict sense,
responsible for his concern about the age of the earth.[58] His interest in
geochronology and his opposition to uniformitarianism were based
upon scientific considerations alone, but he believed that scientific laws
presupposed a regularity in nature which in turn evidenced the working of a providential intelligence.

It must be stressed that Kelvin's belief in the evidence of design in
nature did not entail any belief in a continuous interference in the
mechanism of the universe, or even in an occasional miracle. Quite the
contrary, he asserted repeatedly that, "If a probable solution [to any
scientific problem], consistent with the ordinary course of nature can
be found, we must not invoke an abnormal act of creative power."[59] It
was only in the creation of life that Kelvin believed a divine alteration in
the laws of nature was either necessary or probable. In all other matters

he believed that the universe follows regular, unvarying laws, and it was just this regularity that makes science possible.

> The course of speculative and physical science is absolutely irresistible as regards the relation between cause and effect. Whenever we can find a possible antecedent condition of matter, we cannot help inferring that possible antecedent did really exist as a preceding condition—a condition, it may be, preceding any historical information we can have—but preceding and being a condition from which the present condition of things had originated by force acting according to the laws controlling all matter.[60]

The laws of nature envisaged by Kelvin were *real* in the universalist sense of the term. They were not merely formulated by scientists; they were discovered by science through investigation and proper reasoning. He was confident, moreover, that these laws were ultimately the work of a creative intelligence and that they governed the universe according to the design of a Creator. This point was fundamental to his belief in science, and yet his own concern, and that of science in general as he defined it, was with the discovery and application of natural laws, not with their origin. The task of science, therefore, was to discover the immediate natural cause for any phenomenon without undue concern for its ultimate cause. And in the case of the age of the earth, he was convinced that this procedure led inexorably to the conclusion that time was of finite and relatively limited duration.

Kelvin's interpretation of the relationship between natural law and divine order was neither novel nor profound. The belief in teleology was ubiquitous in nineteenth century thought, and its importance has been widely recognized and discussed. Even the defenders of natural selection, for all of its emphasis upon chance variation, shared a belief in providential design unfolded through natural law. Darwin may have been thinking in metaphor when he wrote of "the laws impressed on matter by the Creator," but he expressed the feeling of his age.[61] Kelvin also shared the biases of his contemporaries in his search for immediate causes in nature and his disregard for ultimate causes. This was just the kind of explanation that Darwin sought in natural selection, and certainly it was what James Hutton had in mind when he said, "The result, therefore, of our *present enquiry* is, that we find no vestige of a beginning, —no prospect of an end."[62] Lyell of course had gone still further in requiring uniformity in the operation of both physical laws and natural processes throughout all geological time, and in the process had lent some justification to Kelvin's accusation that uniformitarianism had confused the meaning of *laws now existing* with the *present order* and *present conditions* in nature.[63] But in spite of such disagreements, and the distinctions involved were frequently very vague, Kelvin's position was intimately attuned to the bias of the age, even among his opponents; and by repeatedly stressing his teleological views, he enforced rather than detracted from the scientific acceptability of his conclusions about the age of the earth.

Within the general framework of natural law and divine order, the universe which Kelvin sought to understand was mechanical and the laws which governed it were exact. His arguments concerning geological time, particularly those dealing with the original energy of the earth and sun, were developed as necessary consequences of mechanical laws operating upon his model of the primordial universe. His cosmogony was therefore fundamental to his geochronology, and compliance in the latter depended upon an acceptance of the former. His choice of the nebular hypothesis, which reigned almost unchallenged as the cosmogony of Victorian science, assured just such an acceptance. To be sure, the nebular hypothesis was itself necessarily speculative. Its postulated model of a primordial gaseous universe producing stars and planets under the influence of mutual gravitation required the acceptance of a large number of unproven and unprovable assumptions, but the assumptions were both eminently reasonable and impossible to disprove directly. Furthermore, the hypothesis was founded upon Newtonian physics and gravitational astronomy, and it had been formulated by one of the foremost mathematicians and one of the most eminent philosophers of all time. It was, in short, the best and most authoritative cosmogonal theory available. And almost equally important from Kelvin's point of view, it was ideally suited to the kind of mechanical explanations that he believed provided the only accurate understanding of the operations of the universe.

Although the nebular hypothesis assured Kelvin of a general concurrence in one of his major premises, his efforts to extend and quantify its implications required numerous additional assumptions that were not so well established. Among these were the assumptions that the sum of the gravitational potential energies of the individual particles in the primordial nebula constituted the original source of all the energy available in the solar system, that these particles were originally remote from each other, and that they were originally at rest.

Kelvin made no attempt to account for this primordial condition of matter, but thermodynamics and mechanics would enable him to calculate the total energy available in any finite portion of such a universe. His calculations of the heat of the earth and sun and consequently his whole approach to geochronology were the direct result of applying these ideas to two particular mechanical systems.

Kelvin's additions to the nebular theory, like the theory itself, were speculative and beyond proof. But also like the original theory, they seemed reasonable and were perfectly in harmony with current discoveries and contemporary scientific reasoning. They also had the advantages of unity and relative simplicity. The most important of Kelvin's arguments, those dealing with the secular cooling of the earth and sun, assumed that the two bodies had been formed in the same way from the same materials and that they obeyed similar natural laws. The difference between the two lay solely in the fact that the less massive earth had already cooled below its fusion point while the sun remained much hotter and still molten. This treatment followed directly from the nebular theory and the belief in the universal applicability of natural laws, and it agreed well with the new spectrosopic discoveries of earthly elements in the solar spectrum. Just as had been the case in the argument from design, Kelvin's assumptions were not novel. He had explored the problem of the earth's age from a new direction, but he had used the established concepts of mechanical analogy, classical dynamics, energy conversion, and the nebular hypothesis. No one was unaware of the speculative nature of many of his assumptions, but they were generally considered to be among the most probable postulates available to science. They were common tender in Victorian scientific thought, and their very lack of novelty gave credibility, and even authority, to the conclusions drawn from them.

But if belief in the universal validity of natural laws was implicit in every step in Kelvin's reasoning, two laws in particular, the first and second laws of thermodynamics, were by far the most fundamental to his approach to the earth's age. He believed that the proper application of thermodynamics would provide valuable insights into many basic geological problems. Indeed, it was his conviction that the principles of thermodynamics were necessarily inviolable that was largely responsible for his interest in geological dynamics and geochronology. He was also convinced that uniformitarianism had led geology astray by ignoring those principles. Thus, his accusations that contemporary geological theory violated the principles of natural philosophy referred almost always to the conflict between the principle of the dissipation of energy and the uniformitarians' demands for unlimited time and unvarying geological forces.[64] But his strongest criticism was reserved for the geologists' misuse of physical theory itself. Lyell's use of the conservation of energy in order to justify the possibility of a continuous cyclic transfer of energy in the sun, for example, was to Kelvin the worst kind of scientific heresy. He also rejected the analogies which the uniformitarians sometimes made between infinite reaches of space and eternal spans of time. Geological time, he argued, was concerned only with time in a finite part of the universe. The universe itself might be infinite, but that had no real bearing upon the earth's age or upon his arguments. In the finite system composed of the sun, the earth, and the solar system, the amount of available useful energy must be finite; it must be constantly dissipated according to the second law of thermodynamics; and the system must be running down. The past activities of both the earth and sun must have been greater than those going on at present, and they cannot have been repeated in an indefinte succession of cycles. In other words, Kelvin believed that complete uniformity of action in geology violated the second law of thermodynamics and thus could not itself be a law of nature.

But even if the second law were disregarded, the quantitative part of Kelvin's arguments, which relied only on the first law and the principles of classical mechanics, would still permit the calculation of the amount of energy available to the solar system from gravitation alone. The conservation of energy also made it possible to calculate the magnitudes of the probable energy conversions within the stystem and to speculate reasonably about the probable effects of both past and present energy transformations. Kelvin's aim was to extend the limits of this speculation until he could locate, define, and date the original condition of the earth.

No one in England was better qualified than Kelvin to introduce the laws of thermodynamics into cosmology. His scientific ability was unquestioned, and much of his prestige stemmed from his contributions to thermodynamics. He had been the first to recognize the importance of Joule's discovery of the mechanical equivalent of heat and the conservation of energy; and independently of Clausius, he had formulated the principle of the dissipation of energy which he used so effectively against uniformitarianism. It was in great measure through Kelvin's efforts that the laws of thermodynamics had been shown to be effective and versatile tools for scientific speculation. These laws may not have had the general appeal to the layman of teleology or the nebular theory—at least not until late in the century—but among scientists their value had been proven and their validity established. Certainly, if unproven assumptions had to be made in order to apply thermodynamics to cosmology, no one was better qualified than Kelvin to judge the credibility of those assumptions, and he was careful to assign limits to this credibility. And finally, he laid continual stress upon the belief that science must work within the limits of what is known, and insofar as possible, he followed his own advice. Thus given his position in science and the almost universal acceptance of his basic presuppositions, it is not surprising that as long as his particular assumptions were unprovable, the burden of disproof rested, with some justification, upon his opponents.

REFERENCES

[1]Thompson (1910), *Kelvin*, I: 9-10.

[2]*Ibid.*, I: 185-188. Even the title of this dissertation is in some doubt. Thompson recorded it as "De Caloris distributione per terrae corpus," while Kelvin referred to it as "De Motu Caloris per Terrae Corpus." See Kelvin (1882-1911), *Mathematical Papers*, III: 295.

[3]Kelvin (1852), *Dissipation of Mechanical Energy*, p. 514.

[4]Waterston (1853), *Temperature and Mechanical Force*, pp. 11-12.

[5]Joule (1847), *Mechanical Equivalent of Heat*, pp. 173-176.

[6]Kelvin (1854a), *Energies of the Solar System*, pp. 1-25.

[7]Kelvin's arguments here are remarkably similar to those put forward several years earlier by J. R. Mayer. But Mayer's *Beiträge zur Dynamik des Himmels, in popularer Darstellung* (Heilbronn, 1848) had been privately printed and was almost unknown in England until the early sixties when it played a prominent role in the controversy between Kelvin, P. G. Tait, and John Tyndall over the priority of the discovery of the conservation of energy. It finally appeared in English in April 1863 as "On Celestial Dynamics." See Mayer, J. R. (1863a).

[8]Kelvin (1854a), *Energies of the Solar System*, p. 21. Note added in May 1854.

[9]*Ibid.*, pp. 24-25. Note added in August 1854.

[10]Kelvin (1854c), *Mechanical Antecedents*, pp. 37-38.

[11]*Ibid.*, p. 40.

[12]Kelvin (1859b), *Investigations of M. Le Verrier*, pp. 134-137.

[13]Kelvin (1860), *Variations of the Periodic Times*, pp. 138-140.

[14]Thompson (1910), *Kelvin*, I:411-414. King, A. G. (1925), *Kelvin the Man*, p. 100. The

paper was Kelvin (1816b), *Physical Considerations,* pp. 141-144.

[15]Kelvin (1862), *Age of the Sun's Heat,* pp. 370-375.

[16]*Ibid.,* p. 372. Quoted from Kelvin (1854a)*Energies of the Solar System,* p. 5. (His italics)

[17]Kelvin (1862), *Age of the Sun's Heat,* pp. 373-375. The basic principle outlined here applies equally well to the theory that meteors are constantly replenishing the sun's heat, and to the theory that they were merely the source of its primordial energy. It was the growing weight of astronomical evidence showing that the supply of meteors available was inadequate for his hypothesis, not the methodological problems involved, that forced Kelvin to abandon his original position.

[18] Helmholtz (1856), *Interaction of Natural Forces.* The original lecture, "Über die Wechselwirkung der Naturkraft" was delivered on 7 Feb. 1854 in Könisberg, Prussia, Kant's native city. Kant's nebular theory first appeared anonymously in 1755 and Laplace's version some forty years later. See Kant (1755) *Allgemeine Naturgeschichte* and Laplace (1796), *Système du Monde.*

[19]Kelvin (1854c), *Mechanical Antecedents,* pp. 38-39.

[20]King, A. G. (1925), *Kelvin the Man,* pp. 100-102. From a letter from Kelvin to his brother-in-law, David King.

[21]Thompson (1910), *Kelvin,* I: 411-417.

[22]Kelvin (1862), *Age of the Sun's Heat,* p. 375.

[23]*Ibid.,* p. 368.

[24]Thompson (1910), *Kelvin,* I: 539. Letter from Phillips to Kelvin, 12 June 1861.

[25]Between 1850 and 1853 Lyell and Hopkins as successive presidents of the Geological Society of London devoted their presidential addresses to a debate over the relative merits of uniformitarian and catastrophic geology. (See: Lyell (1850a), *Presidential Address*; Lyell (1851), *Presidential Address*; Hopkins (1853). *Presidential Address.* For a discussion of this debate see: Cannon (1960), *Uniformitarian-Catastrophist Debate.*)

[26]Kelvin's failure to keep up with the literature even in his own field was notorious. It led him frequently to repeat experiments and discoveries already in the literature. So it is hardly likely that he ever systematically read Lyell's *Principles* which was both voluminous and largely outside his own major interests.

[27]Kelvin (1854a), *Energies of the Solar System,* pp. 8-9.

[28]Here, as in the case of Lyell's *Principles,* it seems unlikely that Kelvin ever systematically read the *Origin.* Darwin's calculation of time based on the denudation of the Weald (see chapter 3) was widely criticized in both the popular and scientific press, and thus Kelvin had no trouble in finding the point for attack. Certainly it is significant that he continued to attack Darwin's calculation long after it had been removed from subsequent editions of the *Origin.* He was either unaware of or chose to ignore the fact that Darwin had retreated on the question of an exact time scale.

[29]Kelvin (1871c), *Presidential Address,* pp. 197-205.

[30]Kelvin (1863a), *Secular Cooling.*

[31]Kelvin's approach to the problem of the earth's secutar cooling relied heavily upon the elegant mathematical analysis of heat transfer published by J. B. J. Fourier in 1822

(See Fourier (1822), *Théorie analytique de la chaleur.*) and read by Kelvin in 1840. Moreover, Kelvin's first published paper, prepared before he entered Cambridge in 1841, was a defense of part of Fourier's theory, while his subsequent work makes it clear that the problems of heat transfer, including their application to the earth itself, occupied a significant part of his attention over the next several decades. (See Thompson (1910), *Kelvin*, I: 14-22.)

[32] In 1849, Forbes was engaged in a five-year series of observations of temperature gradients in different minerals and at different depths in several locations around Edinburgh. Kelvin apparently assisted him for a while and made use of the data gathered for many years. (See: Thompson (1910), *Kelvin*, I: 210 and Shairp, Tait, and Adams (1873), *Life of Forbes*, pp. 463-464.) Forbes also provided additional data when Kelvin was preparing "On the Secular Cooling of the Earth" in 1861. (See letter from Forbes to J. Phillips in Phillips' (1869), *Vesuvius*, pp. 345-347.) Between those dates, Kelvin published several papers on the problem of underground heat in which many of the elements of his subsequent calculation of time were developed, although the age of the earth itself was not mentioned. (See: Kelvin (1855), *Observations of Terrestrial Temperature;* Kelvin (1859a), *Variations of Underground Temperature;* and Kelvin (1861a), *Observations of Underground Temperature.*)

[33] See especially: Cordier (1827), *Température de l'interieur de la terre;* and Fourier (1827), *Température du globe terrestre.*

[34] Descartes (1824-26), *Le Monde;* Leibniz (1859), *Protogée;* Buffon (1853-54), *Epochs of Nature.*

[35] Hopkins (1837-42), *Interior of the Earth;* Hopkins (1839), *Precession and Nutation.*

[36] Kelvin (1863a), *Secular Cooling,* p. 300.

[37] *Ibid.*, pp. 295-299.

[38] Kelvin (1871a), *Geological Time.* According to S. P. Thompson, this address may have been read for Kelvin, since at about this time his wife's continued ill health carried him to the continent and for many months he read no other papers. (See: Thompson (1910), *Kelvin*, I: 527.)

[39] Kant (1755), *Allgemeine Naturgeschichte;* Mayer (1863a), *Celestial Dynamics;* pp. 403-409. Helmholtz (1856), *Interaction of Natural Forces,* pp. 510-12.

[40] Kelvin (1871a), *Geological Time,* pp. 17-32, 39-40.

[41] *Ibid.*, p. 36.

[42] Kelvin (1866c), *Observations and Calculations,* pp. 337-341.

[43] See, e.g.: Kelvin (1872), *Rigidity of the Earth;* Kelvin (1874), *Geological Changes and the Earth's Rotation;* Kelvin (1876), *Review of the Evidence;* Kelvin (1882), *Internal Condition of the Earth.* (Also see letter from Kelvin to G. H. Darwin, 28 Dec. 1881 in Thompson (1910), *Kelvin*, II: 778-779.)

[44] Kelvin's statements of this belief appear as early as 1854 and continue throughout the century. (See: Kelvin (1854c), *Mechanical Antecedents,* pp. 36-37; Kelvin (1871a), *Geological Time,* pp. 51-52; Kelvin (1871b), *Geological Dynamics,* p.129; Kelvin (1881), *Sources of Energy in Nature,* pp. 435-37; Kelvin (1889), *On the Sun's Heat,* pp. 378-379, 391; and Kelvin (1892a), *Dissipation of Energy,* pp. 473-74.)

[45] Thompson (1910), *Kelvin*, II: 860.

[46]Kelvin (1889), *On the Sun's Heat.*

[47]*Ibid.*, p. 397.

[48]Kelvin (1892a), *Dissipation of Energy,* p. 474. The quotation is taken from Kelvin (1852), *Dissipation of Mechanical Energy,* p. 514. (My italics)

[49]Kelvin (1863a), *Secular Cooling,* p. 300; Kelvin (1871a), *Geological Time,* p. 64; Kelvin (1876), *Review of the Evidence,* p. 243; Kelvin (1899), *Age of the Earth,* pp. 215-216, 226; Thompson (1910), *Kelvin,* II: 779.

[50]Kelvin (1895a), *Age of the Earth,* p. 227. The full importance of the exchange with Perry is discussed in Chapter 5.

[51]Kelvin (1876), *Review of the Evidence,* pp. 238-43.

[52]Kelvin never published an endorsement of the implications of radioactivity for geological time. On the contrary, as is discussed in Chapter 6, he opposed the idea that radioactive materials could spontaneously emit heat without being supplied by an outside source. Nonetheless, J. J. Thomson reports that in private conversation Kelvin did concede that his theories had been overthrown. (See: Thomson, J. J. (1936), *Recollections,* p. 420.)

[53]Archibald Geikie attributed to Joseph Larmor the following statement, supposedly uttered at Kelvin's funeral: "Conceive a perfectly level line drawn from the summit of Newton's genius across all the intervening generations; probably the only man who reached it in these two centuries has been Kelvin." Geikie, A. (1924), *Autobiography,* pp. 350-351. Such opinions, if not quite so extreme or poetic, were common, ranging from Helmholtz's opinion of the young Kelvin (Thompson (1910), *Kelvin,* I:310, 324-25) to Frank Harris' observation on the leader of British science (Harris (1963), *Life and Loves,* p. 389).

[54]Quoted in King, A. G. (1925), *Kelvin the Man,* p. 96.

[55]A brief but authoritative discussion of this point is given in Thomson, J. J. (1936), *Recollections,* p. 421.

[56]Perry (1895a), *Age of the Earth,* p. 224.

[57]Jenkin (1867), *Origin of Species.*

[58]Thompson (1910), *Kelvin,* II: 1086-1097; King, A. G. (1925), *Kelvin the Man,* pp. 28-31. This opinion, it should be noted, is directly contrary to that expressed in Eiseley (1961), *Darwin's Century,* pp. 234-235.

[59]Kelvin (1871c), *Presidential Address,* p. 200.

[60]Kelvin (1871a), *Geological Time,* p. 35.

[61]Darwin, C. (1859), *Origin,* p. 488, and all subsequent editions. (Also see: Mandelbaum (1958), *Darwin's Religious Views.*)

[62]Hutton (1788), *Theory of the Earth,* p. 304.

[63]For a valuable discussion of Lyell's views on this point see: Rudwick (1970), *Strategy of Lyell's Principles,* especially pp. 7-8.

[64]Kelvin makes this point specifically clear several times, as for example in Kelvin (1863a), *Secular Cooling,* p. 295 and Kelvin (1871b), *Geological Dynamics,* p. 77.

III
Kelvin's Influence:
The Initial Reception

Speaking before the geological section of the British Association in 1899, Archibald Geikie reminded his audience of how geologists in the 1860's had been "startled by a bold irruption into their camp from the side of physics."[1] He referred, of course, to the publication of Kelvin's papers on the age of the earth. But time had telescoped events in his memory, and his recollections were somewhat distorted. By 1899 Kelvin's influence had produced a major change in geological thought. The more radical implications of uniformitarianism had been left behind, and the infinite or vaguely indefinite time scales of the mid-century had given way to an earth of finite and calculable antiquity. By 1899, indeed, Kelvin's original estimate of 100 million years for the earth's age had become so entrenched among geologists that they were sharply at odds with his more restrictive later results. But such changes had hardly been sudden. Far from taking geology by storm, Kelvin's chronology had been adopted gradually by individual scientists, and nearly a decade passed before its full impact began to be felt.

A survey of British geological literature in the sixties is much more revealing in the general absence of references to the age of the earth than in the presence of the few papers where it was discussed. References to geological time were almost invariably restricted to the relative age of one system of strata with respect to other geological periods or epochs. The only question of chronology that stood out was the age of man. The discovery of the remains of prehistoric men and the debates over Darwinian evolution aroused considerable interest in man's own antiquity; but again most of the chronologies produced were relative, and few ventured to express the age of man in years.

Kelvin was not completely ignored, but initially the geologists of the sixties seemed more interested in his discussion of the earth's rigidity than his chronology. Debates over whether the earth's interior was solid or semi-fluid had been simmering since William Hopkins announced his conclusion that a rigid core was absolutely necessary to account for the earth's axial precession and nutation.[2] For most geologists, however, Hopkins' physical and astronomical arguments were less convincing than the fact that a molten interior seemed necessary to account for the geological phenomena of volcanoes, earthquakes, mountain formation, and other evidence of massive shifts in the earth's surface. Thus in extending Hopkins' arguments, Kelvin aroused new interest in the state of the earth's interior, but in the process diverted attention from his chronology.[3]

Despite Kelvin's criticism, uniformitarianism became more entrenched than ever during the early sixties. Darwin in particular elevated the uniformitarian conception of time to a new prominence as a critical part of the theory of natural selection, and Lyell remained the recognized leader of theoretical geology. In fact "On the Sun's Heat" and "On the Secular Cooling of the Earth" seem to have had no more immediate effect than the papers of 1854.[4] There was still no outcry from geology and no significant change in the uniformitarian position. Kelvin's frustration mounted, and in 1865 it burst forth in a direct challenge to the geologists:

> The "Doctrine of Uniformity" in Geology, as held by many of the most eminent of British Geologists, assumes that the earth's surface and upper crust have been nearly as they are at present in temperature, and other physical qualities, during millions of millions of years. But the heat which we know by observation, to be now conducted out of the earth yearly is so great that if this action had been going on with any approach to uniformity for 20,000 million years, the amount of heat lost out of the earth would have been about as much as would heat by 100° Cent., a quantity of ordinary surface rock of 100 times the earth's bulk. (. . .) This would be more than enough to melt a mass of surface rock equal in bulk to the *whole earth*. No hypothesis as to chemical action, internal fluidity, effects of pressure at great depth, or possible character of the substances in the interior of the earth, possessing the smallest vestige of probability, can justify the supposition that the earth's upper crust has remained nearly as it is, while from the whole, or from any part, of the earth, so great a quantity of heat has been lost.[5]

Kelvin's work had not gone entirely unnoticed, however. By the end of the decade a few geologists had begun to accept his conclusions, and even such notable opponents as Darwin, Huxley, and Lyell had felt obliged to make some response. But perhaps even more significant in the long run, a number of geologists had been led to look at geology quantitatively. They began to look beyond the mere ratios of relative time scales and to seek ways of measuring the rates of geological change. It was perhaps a delayed beginning, but once started the change came quickly.

Converts and Opponents

The first important quantitative determination of the earth's age based exclusively on geological data was published by John Phillips in 1860, nearly two years before Kelvin's results appeared. Phillips had been raised by his uncle, William Smith, the founder of stratigraphy, and as a willing pupil had accompanied Smith on his wanderings. It is hardly surprising, therefore, that he looked upon stratigraphy as the key to understanding geology, and that he looked askance at some of the vagueness inherent in uniformitarianism. As early as 1837, for example, he questioned the validity of Hutton's "no trace of a beginning, no prospect of an end," and asserted that time was a necessary element in geological explanation. He even went so far as to propose that "the total length of the scale of strata was of importance as an element for the direct computation of the total time elapsed in the formation of the crust of the globe."[6] But it was not until spurred by the extravagance of Darwin's 300 million years for the denudation of Weald, that he attempted such a computation of his own.

Phillips' calculation pioneered the use of several assumptions and procedures that were to be widely used in later calculations of the earth's age. First of all, although not a uniformitarian and frankly opposed to the idea that the rates of geological activity are constant even in the present, let alone that they could be considered uniform for all past time, he accepted the practical necessity of assuming uniformity for the sake of calculation. Secondly, following up his own suggestion of 1837, he adopted the thickness of sedimentary strata as the most direct measure of past time. Thus the simple ratio of the total thickness of sedimentary rocks to the constant rate of average sediment formation should provide a direct calculation of the age of the earth's crust. Such a calculation required numerous additional assumptions, however. Estimates were necessary to determine the rate at which sediment

is produced by the erosion of existing formations, and some relationship had to be established between the area of land eroded to provide
the sediment and the area of the ocean bottom receiving it. For the sake
of calculation, Phillips assumed a one-to-one ratio for the latter case,
and for the former, chose as representative of the world as a whole a
selected area for which data was available, the Ganges River Basin. He
then assumed that the strata of the river basin had been laid down at
the same rate at which it is presently being eroded away, and calculated
that 95,904,000 years had been required for its formation.[7]

Phillips placed little credence in the precision of his calculation. The
Ganges, after all, carries an unusually heavy load of sediment, and he
recognized that any calculation based on it alone would probably
underestimate the earth's age. On the other hand, he believed it very
probable that the area of sedimentary accumulation was actually smaller than the area of concurrent denudation, and so his result might be
too high. In short, the range of possible error was vast, and the assumption of uniformity itself introduced far too much uncertainty for Phillips to regard his results as in any way exact. As he cautioned his
readers, "In whatever way we try the question of the antiquity of the
fossiliferous strata . . . the results are always the same, always indicative
of periods too vast, and we must add too vague, for conception."[8] And
yet there were limits even to imprecision, and Phillips still confidently
rejected Darwin's excessively high result on the one hand and one of
his own corrections which would limit geological time to an unaccept-
ably low 38 million years on the other. His various auxiliary calculations
based on data from several different formations did, after all, agree in
order of magnitude, and he regarded that as highly significant. Thus
whatever the uncertainties, Phillips' result was a far cry from the
unlimited or indefinite time of Lyell.

Phillips' work had little immediate effect upon his fellow geologists.
His chronology was buried in a book on the development of life and
was eclipsed by the generally anti-Darwinian arguments of the work as
a whole. Even Kelvin, although well aware that Phillips' results agreed
remarkably with his own conclusions from the earth's secular cooling,
made no mention of them for several years.[9] This oversight seems
particularly odd since Phillips was among the first to support Kelvin's
arguments publicly, even while entertaining reservations about their
complete accuracy.[10] In fact, as we shall see later, the only person who
was really struck by Phillips' argument was Darwin himself.

A few other people had begun to recognize the significance of
Kelvin's arguments by the mid-sixties, but in his own opinion the first

important convert was Archibald Geikie (1835-1924), the newly appointed director of the Geological Survey for Scotland. Geikie was a
fellow Scot and personal friend, and by 1867 had already begun to
demonstrate the talents that were to make him one of geology's most
influential public spokesmen. At the time, however, he had no real
interest in geochronology, and in fact specifically denied that the
geological record could provide the data for computing the earth's age.
If calculations of geological time in years were ever to be possible at all,
he believed that the data would probably come from astronomy.[11] A
year later that belief led him firmly into Kelvin's camp.

In a paper delivered in 1868 Geikie carefully surveyed the information available concerning the nature, rates, and extent of denudation
throughout the world.[12] The results were admittedly uneven and
incomplete, but he believed that the best of the quantitative data could
support some meaningful generalizations. Among other things, he
observed that the continents are presently being eroded away at such a
rate that they would all be leveled within a few million years if new ones
were not concurrently raised from the sea beds. Furthermore, although he had no clear evidence to support him, he believed that if the
present rates of geological activity did not represent the average for all
time, they were more probably below the mean than above it. The
conclusion was obvious: geologists had grossly underestimated the
rates of present and past geological activity, and those who had used
denudation as evidence for the inconceivably slow progress of geological change were wrong. Thus, he asserted, "our demands for enormous periods of past time, insofar as based on the evidence of past
denudation, are unnecessary." Geikie was willing to concede to the
uniformitarians the point that since measurements of the present rates
of denudation are the only ones available, they could justifiably be used
as measurements for the past so long as they were in accord with the
evidence. But he did not himself undertake any calculation of time
from the data accumulated. Instead, almost as an afterthought, he
cited the "recent researches in physics" which showed "that the unlimited ages demanded by geologists cannot be granted," and then uncritically accepted the conclusion of Kelvin and James Croll that "About
100 millions of years is the time assigned within which all geological
history must be comprised."[13]

The geological data which Geikie had accumulated did not in themselves indicate a restrictive geochronology, certainly not 100 million
years. They *could* yield such a calculation if the necessary assumptions
were made, but the assumptions were in no way implicit in the data.

Geikie, however, was candidly willing to accept the accuracy of the physicist's arguments, and because of them to abandon the uniformitarian concept of geological time. In the process, he had to re-evaluate the prevailing ideas about the relationship between past and present geological actions, and in so doing, he was the harbinger of a new interpretation of uniformitarianism.

For Geikie, and for a growing number of geologists after him, uniformitarianism remained preferable to catastrophism because it was based upon experience, but no longer could the premises of uniformity be justifiably pushed to extreme. The present could be accepted as the key to the past only so long as the evidence permitted, and he, repeatedly warned against dogmatically assuming that no change in geological activity had occurred. Uniformitarianism, in this sense, meant that the geological agents responsible for past changes in the earth's crust were the same as those now acting, but it did not necessarily mean that the present rate of activity was the same as past rates. Uniformity of *kind* was accepted but not uniformity of *degree*. The new uniformitarianism still upheld the value of empirical study and of induction from present geological processes, but sought to moderate geochronological speculation and to reconcile geological evidence with a shortened time scale.

Although Geikie was the first geologist whom Kelvin acknowledged as accepting his estimate of the earth's age, another Scot, James Croll, had begun several years earlier to develop the implications of the new geochronology and had in fact gone far beyond Geikie in his detailed analysis of the problem. Kelvin's initial oversight was understandable, however. Croll (1821-1890) was to become one of the most influential advocates of the limited time scale, but in the mid-sixties he was a middle-aged, self-educated unknown.. Indeed, when he first announced his advocacy of Kelvin's chronology in 1864,[14] he still occupied the menial post of keeper at the Andersonian Museum in Glasgow. His geological papers soon attracted attention, however, including that of Geikie and A. C. Ramsay, the director of the Geological Survey of England, and in 1867 they had him appointed to the Scottish Survey. He thereafter came into contact with many of the best geologists of the age, and during the next decade built a considerable reputation among British scientists with his studies on the causes of glaciation and the geochronology that he developed from them.[15]

Croll never considered himself to be a geologist. He had taught himself physics during his years at the Andersonian, but admitted to having little knowledge of geology and no great liking for it except as it related to physical questions. He consequently avoided the biases of more conventional geologists, and was candidly willing to throw out geological estimates of time when they conflicted with the results of physics.[16] He was convinced, moreover, that uniformitarian geologists had grossly overestimated the length of time required by their data because of their inability to conceive of vast periods of time in a true perspective. As he put the question:

> But is it the case that geology really requires such enormous periods as is generally supposed? At present, geological estimates of time are little else than mere conjectures. Geological science has hither to afforded no trustworthy means of estimating the positive length of geological epochs. Geological phenomena tell us most emphatically that these periods must be long; but how long, these phenomena have, as yet, failed to inform us . . . . They present to the eye, as it were, a sensuous representation of time; the mind thus becomes deeply impressed with a sense of immense duration; and when one under these feelings is called upon to put down in figures what he believes will represent that duration, he is very apt to be deceived.[17]

Both geology and physics, Croll argued, opposed the idea of an unlimited age of the earth, and both offered good reasons for limiting its age to less than 100 million years.

Although impressed by Kelvin's chronological arguments, Croll did not accept them blindly. He adopted the argument from the earth's secular cooling and the 100 million year limit on the earth's age, but expressed serious reservations about the question of the sun's heat, and rejected outright the argument from tidal retardation. In the latter case he admitted that the earth would have solidified with a pronounced bulge at the equator if its rotational velocity at the time of solidification were much greater than it is now. But calculations based upon the most recent methods of measuring rates of denudation indicated that any equitorial bulge that may have existed would have eroded away as the sea levels adjusted to the velocity of the earth's rotation. Thus Croll argued that even if the earth is perfectly rigid, its present shape can give no indication of its past configuration, nor can its assumed rotational deceleration give any measure of its age.[18]

The limitation of the sun's age presented a different kind of problem, for though Croll accepted the fundamentals of Kelvin's argument, he believed that the result provided too short a time span to be reconciled with the result from the earth's secular cooling. Like Kelvin, Croll believed that the sun's heat could only be explained mechanically, but gravitational energy alone could account for no more than 33 million years' supply of solar energy, and the Kelvin-Helmholtz hypothesis allowed for only 20 million years. Croll, however, needed to account for 100 million. Chemical energy was even less adequate. Croll's solution, therefore, was to add a new element to Kelvin's calculation. Instead of assuming that the primordial particles from which the sun had formed were originally cold, he assumed that they were hot. This condition could have come about, he suggested, if the original solar nebula had itself been formed from the collision of two celestial bodies. He made no attempt to acccount for the origin or initial motion of these bodies other than to assert that their motion was not necessarily limited by the force of their mutual attraction. He did suggest, however, that the heat generated by the collision would account for the dispersion of matter into the nebula that had formed the starting point for Kelvin's hypothesis. And of course, it would be an additional source of energy for the sun itself.[19]

Croll's solar hypothesis had two advantages over Kelvin's. It pushed the causal chain of mechanical explanation, and hence the question of ultimate origins, one step further into the remote past. But more important, since the velocities of his pre-nebular celestial bodies were unknown and unknowable, they could account for almost any duration of solar heat. The energy available would necessarily be finite so it provided no support for radical uniformitarianism, but it was not limited by gravitation alone and thus could easily account for the 100 million years that Croll needed to reconcile the results of Kelvin's arguments. A major difficulty, on the other hand, was that the hypothesis introduced yet another unprovable assumption which had the net effect of making meaningful quantitative results impossible. Another serious problem might have arisen from Croll's method of treating the macroscopic meteoric particles of the nebula as if they followed the same laws of compression as the microscopic particles of a gas, but in the infancy of the dynamic theory of heat such gross mechanical analogies were common and this one passed unnoticed. Thus on balance, Croll believed he had created an hypothesis that would satisfy both physics and geology once the data were properly interpreted.

Although Croll professed to having little interest in geological time itself, his attempted reconciliation of Kelvin's results from the heat of the earth and sun was but a minor part of his contribution to geochronology. Of much greater influence was his creation of an entirely new method of measuring the earth's age. Characteristically, his approach to the question of time arose indirectly out of his concern for the mutual relationships between astronomy, physics, and geology, and indeed appeared first as a mere corollary to his proposed explanation of the cause of the great ice ages. The circumstance was fortunate, for during the sixties interest in glaciers and the ice ages was near its peak in Britain. And consequently Croll found both an audience for his hypothesis and his first professional position.

The basis for Croll's glacial theory was the fact, long known to astronomers, that the gravitational pull of the other planets in the solar system constantly alters the shape of the earth's orbit, causing it to oscillate between near circularity and pronounced ellipticity. Such changes would necessarily affect the relative lengths of the seasons, and Croll's suggestion was that during periods of high orbital eccentricity (that is, when the elliptical shape of the orbit is most pronounced) the hemisphere experiencing winter at the aphelion of the orbit would suffer long harsh winters whose effects would be only partially offset by the short summers. There should consequently be a gradual accumulation of ice and snow in one hemisphere while the other enjoyed a moderate climate produced by long pleasant summers and short mild winters. Croll further postulated that because of the precession of the equinoxes these extremes in climate would alternate from one hemisphere to the other every 10,500 years.

Thus far Croll's hypothesis was not original. Although neither he nor many of his contemporaries were aware of it at the time, a very similar proposal had been made in 1842 by a French mathematician, Alphonse Joseph Adhémar. And indeed, similar, though less specific, discussions of the possible relationship between climatic change and the eccentricity of the earth's orbit had been made and rejected at various times since 1830 by John Herschel, Francois Arago, and Alexander von Humboldt. Herschel, however, had shown that the total heat received by the earth depends only upon the minor axis of its orbit and had concluded that its position in the orbit has little bearing upon the net amount of heat received in either summer or winter.[20] Consequently the earlier versions of the hypothesis were laid aside or passed unnoticed.

When he heard of the work of his predecessors, Croll concluded that they had erred by placing their emphasis exclusively upon the direct effects of astronomical change. In the mature version of his own hypothesis, the astronomical conditions provided only the starting point. He argued that once a climatic change had begun as a result of variations in orbital eccentricity, several meteorological conditions would be induced that would greatly magnify the overall effect. Among other things, he suggested probable shifts in the trade winds that would deflect tropical currents away from the colder hemisphere, and an increase in precipitation in the colder hemisphere which would build up the snow cover. Moreover, the snow itself would serve to deflect the sun's rays and absorb little heat, while dense fogs would shield the earth still more from solar radiation. All in all, Croll believed that the combination of these conditions would be more than enough to account for a glacial age. And since he was equally sure that the other conditions were dependent upon high orbital eccentricity, he was confident that the ice ages could be dated astronomically.[21]

By 1867 he had begun to gather the data necessary to prove his point. Utilizing formulas developed by Leverrier, he calculated orbital eccentricities at regular intervals for the past million years and found a succession of three periods of extremely high eccentricity between 950,000 and 750,000 years ago. He also found several other periods of slightly less pronounced eccentricity at later times, the two most recent having occured approximately 200,000 and 100,000 years ago. His problem therefore was to determine which of these periods corresponded to the ice age that geologists were then so busily studying. And for that he turned to geology.[22]

Initially Croll assumed that the last great ice age had occurred during the period of highest eccentricity which had ended about 700,000 years ago. But even at the time he could not help wondering whether denudation acting over such an enormous period would not have obliterated all trace of glaciation.[23] By 1868 he was sure. In 700,000 years several hundred feet of the earth's surface must have been worn away, more than enough to destroy any record of glacial activity. Equally telling from Croll's point of view, however, was Lyell's calculation, based upon the number of extinct paleontological specimens, that if the last ice age occurred 850,000 years ago, the date of the beginning of the Cambrian must extend back nearly 240 million years. Such a conclusion was out of the question since Croll had already accepted Kelvin's argument that 100 million years was all of the time available for the earth's entire history. If, on the other hand, the most recent ice age were assumed to correspond to the last period of relatively high eccentricity which ended only 80,000 years ago, and if certain necessary changes were made in Lyell's method of calculation, the time since the onset of the Cambrian would be reduced to 60 million years. This estimate, Croll pointed out, was in good agreement with Kelvin's and was consequently the one he adopted.[24]

The completed statement of Croll's hypothesis appeared in 1868, the same year as Geikie's address on the effects of denudation and Kelvin's "On Geological Time," and like them it candidly asserted the necessity of fitting geological data into the time frame established by physics. But Croll's rejection of uniformitarianism, which took up a large part of the paper, had little in common with the blunt frontal attack of Kelvin. Instead he argued that contemporary geologists had been misled by their inability to conceive of vast periods of time and had thus been led to underestimate greatly the rate of action of geological forces. His talent for constructing simple but effective analogies greatly added to the persuasive force of his argument, and coupled with an array of quantitative data drawn from both physics and geology it made his point convincing, especially among those who had just begun to look at geology quantitatively.

Judged in terms of their subsequent influence, Geikie and Croll were important converts to the new chronology, but in 1868 the uniformitarian view of time seemed far from abandoned. After nearly forty years uniformitarianism dominated British geology more completely than ever before, and references to indefinite and even infinite time spans still appeared regularly in the literature.[25] For the rank and file of geologists, time was accepted uncritically with little regard for its exact duration. Nonetheless, Lyell could not completely disregard the arguments against him, and in the tenth edition of his *Principles of Geology* he attempted to answer Kelvin and Croll directly.

Lyell first heard of Croll's glacial hypothesis in 1865, and though he disagreed with much of it, he believed that certain parts demanded attention.[26] His own explanation of the glacial epochs relied heavily upon assumed changes in the distribution of land over the earth's surface. An accumulation of land near the equator would produce a uniformly mild climate, while an accumulation near the poles would create meteorological conditions sufficient to produce widespread glaciation. He looked upon Croll's theory as an auxiliary factor which might intensify the effects of already existing geological conditions. Pronounced orbital eccentricity alone would not cause an ice age, but its importance as a secondary cause might be sufficient to make it a necessary condition for their occurrence. And in that case tables of orbital eccentricity might in fact serve as geological timekeepers.

Lyell's attempt to date the ice ages began well before the final version of Croll's theory appeared, and he consequently had to provide his own modified version of Croll's table of orbital eccentricities. Nonetheless the periods of high eccentricity which he thought might correspond to the glacial epochs were essentially the same ones adopted by Croll only grouped somewhat differently. The reactions of the two men were quite different, however. Lyell immediately rejected the two most recent possibilities, 100,000 and 200,000 years ago respectively, as being too recent to allow for the geological changes that have occured in the post-glacial period, and adopted the period of extreme eccentricity extending from 750,000 to 850,000 years ago. (Croll initially made the same choice, but with serious reservations.) He then discounted the probability that these dates were at all positive, but suggested that they might still give some measure of time. The remainder of his argument was paleontological. Estimating from the available evidence that about ninety-five percent of the fossil mollusca in post-glacial strata are indentical with presently existing species, and assuming that about one million years have elapsed since the beginning of the ice age, he inferred that a complete *revolution* of species would take 20 million years. According to his count, moreover, twelve complete revolutions of species have occurred since the beginning of the Cambrian. Thus, making the probable assumption that each revolution of species required approximately equal time, he concluded that the elapsed time since the beginning of the Cambrian—and hence the total age of existing strata—must be about 240 million years.[27]

In choosing the earlier rather than the later period of eccentricity as the probable beginning of the ice age, Lyell remained true to his basic belief in the extreme slowness of geological processes. But even in entertaining the possibility that 240 million years might be sufficient to account for the formation of existing strata he had broken sharply with his earlier reliance upon indefinite time. Croll, however, still found the result too vast to be commensurate with Kelvin's physics and thus opted for the more recent periods of high eccentricity. His arguments convinced even Lyell—almost. In the eleventh edition of the *Principles* in 1872 Lyell adopted a compromise 200,000 years as the probable date of the ice age, but refused to accept the shortened time scale that this change entailed. Instead, he re-emphasized his earlier insistence upon the geographical causes of glaciation and abandoned any attempt to calculate geological time in years.[28]

Lyell's response to Kelvin was considerably less direct than his treatment of Croll even though the questions involved were far more critical. The main thrust of Croll's argument, after all, had been to show that the present rates of geological processes could accomplish much more in a limited time than the uniformitarians had believed possible. Kelvin on the other hand sought to undermine the very foundations of uniformitarianism itself. If his inferences from the laws of thermodynamics were true, the dynamically balanced, steady-state earth of the uniformitarians could not possibly exist. The intensities of geological forces must have diminished with time, and thus uniformity could not be a law of nature. Lyell was well aware of the dilemma he faced, but could find no adequate solution. His response was consequently piecemeal, and the most important questions were left unanswered.

As early as the eighth edition of the *Principles* in 1850, long before Kelvin's arguments were developed, Lyell had attempted to account for the earth's internal heat by postulating a continuous cyclic transfer of chemical and electrical energy which would generate heat without diminishing the net magnitude of either.[29] In retrospect this idea appears amazingly naïve, and certainly it was based upon an erroneous conception of the conservation of energy; but before the formulation of the second law of thermodynamics in 1852, it was not without some degree of legitimacy. When the same argument was repeated in the tenth edition of 1866-68, however, Lyell was knowingly contradicting the principle of the dissipation of energy. He even went so far as to admit that he might be postulating a perpetual motion machine, but sought justification in the nineteenth century's last line of defense. He could see no reason, he asserted, why science should "despair of detecting proofs of such a regenerating and self-sustaining power in the works of a Divine Artificer."[30] Only a few pages further on, however, it was clear that this explanation was unsatisfactory even to Lyell himself. Citing the recent discovery that the sun's spots influence terrestrial magnetism, he speculated that the earth's heat might be continually restored if its internal electricity and magnetism were somehow drawing energy from the sun.[31] This would eliminate the need for perpetual motion on earth, but only at the expense of transferring the basic problem to the sun. Neither of Lyell's proposals actually came to grips with Kelvin's objections, and both raised serious physical problems which he wisely refrained from elaborating upon. But both were left in all later editions of the *Principles*.

Both Kelvin and Lyell sought above all else to explain the operations of the earth in terms of ascertainable natural laws. It is ironic, therefore, that in order to counter Kelvin's objections to uniformitarianism, Lyell found it necessary to invoke the possibility of divine laws at variance with the discovered laws of nature. He was, in fact, grasping for any possible mechanism which might provide the earth with a continuous and uniform supply of energy; and in that frustrating task he was soon joined by his friend and disciple Charles Darwin.

Kelvin, Darwin, and Wallace

Charles Darwin's (1809-1882) theory of evolution by natural selection was the most famous, the most popularly discussed, and perhaps the most important scientific hypothesis of the nineteenth century. It explained the progressive modification of species in terms of minute chance variations in individuals, reinforced by slow degrees over vast periods of time. The limitation of geological time was consequently of great importance to evolution. If Kelvin's arguments proved valid, they denied the possibility that the earth could have existed long enough to meet the demands of natural selection, and no one was better aware of the problem than Darwin himself. Indeed, unable to cope with the physicist's mathematics, Darwin found himself forced into an awkward and reluctant retreat.

Darwin had been profoundly influenced by Lyell's *Principles of Geology*, and when he proposed his theory of evolution by natural selection in 1859, he was supremely confident that he could count on the world having endured for as long as his theory demanded. Lyell had taught him the importance of time in accounting for geological change—and by analogy for biological change—and had convinced him that the time available for those changes had been inconceivably vast. "He who can read Sir Charles Lyell's grand work on the Principles of Geology," Darwin challenged, "and yet does not admit how incomprehensibly vast have been the past periods of time, may at once close this volume."[32] This challenge appeared in every edition of *On the Origin of Species,* but although few accepted the invitation to close the book, not a few took the opportunity to answer the challenge.

The opening for the debate came in Darwin's single quantitative statement about geological time. Confident in his faith in Lyell's doctrine, he abandoned his usual caution and instead of the meticulous amassing of evidence which characterized the *Origin* as a whole, he chose to illustrate the vastness of time by a single example drawn from a familiar locale, the denudation of the Weald. His approach was simplicity itself. Looking at the great width and depth of the Weald, the great eroded valley stretching between the North and South Downs across the south of England, he compared the total volume of material which must have been eroded away during its formation with a rough estimate of the rate at which marine denudation would wash it away. He concluded that the process must have required about 300 million years.[33] It seems clear that Darwin did not think it necessary to prove the vastness of time. Lyell, after all, had already provided such arguments in abundance. Thus, rather than a careful calculation based on the available quantitative evidence, he contented himself with a naïvely conceived, hasty calculation purely for illustration. It was a mistake he came quickly to regret.

The reaction was immediate. On December 24, 1859, less than a month after the official appearance of the *Origin*, the *Saturday Review* carried a critical review which singled out the calculation of the denudation of the Weald as the focus for attack.[34] Shortly afterwards John Phillips, who had frequently disagreed with Lyell and Darwin in the past, launched his challenge before the Geological Society of London and later included it in his book, *Life on Earth*.[35] But Darwin had already become disconcerted by his unfortunate lapse of caution. Indeed when the second edition of the *Origin* appeared only two days after the criticism in the *Saturday Review*, it already contained a statement that the calculation of the denudation of the Weald might need to be reduced by a factor of two or three.[36] Nonetheless, the reviews stung, and by early January 1860 the calculation had become "those confounded millions of years (not that I think it is probably wrong)."[37] Even the parenthetical reservation did not last the year, however, and in November Darwin wrote Lyell of the changes to be made in the third edition: "The confounded Wealden Calculation to be struck out, and a note to be inserted to the effect that I am convinced of its inaccuracy from a review in the Saturday Review and from Phillips, as I see in his Table of Contents that he alludes to it."[38] The concession hurt bitterly, for a few days later he again wrote Lyell: "Having burned my fingers so consumedly with the Wealden, I am fearful for you . . . for heaven's sake take care of your fingers: to burn them severely, as I have done, is very unpleasant."[39] When the third edition of the *Origin* appeared in early 1861 the Wealden calculation was conspicuously missing, but it was too late. Darwin had set up a straw man and his opponents attacked it joyously.

It was almost a year after the appearance of the third edition of the *Origin* that Kelvin first attacked Darwin's chronology. In all probability he was unaware of Darwin's retreat; but even had he known, it is unlikely to have made any difference in his attitude. The Wealden calculation was gone, but not the necessity for vast periods of time to accommodate natural selection. And Kelvin was opposed to natural selection, as he was to uniformitarianism, because he believed it to be directly contrary to the fundamental principles of nature. His objections were actually threefold: natural selection could not account for the origin of life; it required far more time for its operation than the laws of physics would allow; and its emphasis upon chance did not allow for the evidence of design in nature. Of these objections, Kelvin considered the third to be the most important, but only the second was concrete and subject to scientific verification. It was consequently this objection that he stressed in his attacks upon Darwin's theory.

It should be noted that Kelvin did not object to evolution *per se*, but to the mechanism of natural selection. He was a firm believer that science should seek recourse in a first cause (or divine creation) only after every possibility for natural explanation had been exhausted, and he recognized that evolution could account for the diversity of life by natural means. He was evolutionist enough, in fact, to propose publicly that the primitive seeds of life on earth may have first arrived upon the meteoric fragments of some previously inhabited world and subsequently evolved into forms now existing.[40] Kelvin seems to have feared, however, that the Darwinians would attempt to push their theory to the origins of life itself, and this he was anxious to deny them. Indeed, he denied the validity of such speculations on any scientific grounds whatever.

> I need scarcely say that the beginning and maintenance of life on earth is absolutely and infinitely beyond the range of sound speculation in dynamical science. The only contribution of dynamics to theoretical biology is the absolute negation of automatic commencement or automatic maintenance of life.[41]

On this point, at least, Kelvin could have had Darwin himself as an ally, and thus his strongest objections were directed at the emphasis placed upon chance in natural selection and the implicit denial of providential guidance. Life may have originated in a simple primitive germ, but Kelvin was convinced that the complexity of life bore witness to the work of a Creative Intelligence. He was equally convinced that whereas natural selection would require almost endless time, divine guidance would enable evolution to produce the diversity of life in a relatively short period. Thus as far as evolution was concerned, his arguments for limiting the earth's age were also proofs of design in nature.

The first important reference to the basic incompatibility between Darwin's theory and Kelvin's geochronology appeared in Fleeming Jenkin's famous review of the *Origin* in 1867.[42] Jenkin (1833-85) had much in common with Kelvin. He was a Scot, a physicist, an engineer, and an inventor, and at the time of his review of the *Origin*, the two men were working in close association. They had served together on several transatlantic cable laying enterprises, and as early as 1860 had created a partnership to market a patent. In 1865, the partnership was extended to include the joint marketing of all their patents, and a second was formed as consulting engineers.[43] The partners were more than successful business men, however; they were close personal friends. Thus Jenkin, who clearly looked upon Kelvin as the senior partner (he once referred to himself as a "great worshipper" of Kelvin[44]), was in a position to be thoroughly familiar with the older man's views on geological time, and it is hardly surprising that he used them with such telling effect in his review of the *Origin*.

Jenkin's criticism of the Darwinian theory hinged upon his contention that it was mathematically impossible for a single fortuitous variation in an individual organism to be perpetuated. Any individual variation, he believed, would be swamped in succeeding generations by the preponderance of normal individuals. The inadequacy of geological time was a secondary point in his criticism, but he stressed the value of Kelvin's work as strong reinforcement for his primary argument. Jenkin knew and candidly admitted that Kelvin's results were mere approximations. He conceded that some of the information upon which they were based was totally wrong, and that new data might materially alter Kelvin's quantitative conclusions. Nonetheless, he accepted the basic argument as valid and insisted that it must therefore limit the uniformitarian demands for time. He was convinced, moreover, that whatever quantitative changes had to be made in

Kelvin's results, the maximum time available since the formation of the earth's crust would prove inadequate for Darwin's theory.[45]

In his discussion of geological time, Jenkin's views were very similar to Kelvin's. He saw the principle of the dissipation of energy as absolute proof against the possibility of geological uniformity. The geological activity of the earth must diminish over long periods of time, and thus the present rates of geological processes cannot in themselves provide an accurate measure of past time. Jenkin did not argue for catastrophes in the old sense any more than Kelvin had, but because the rates of geological change could not be constant some other measure of geological time had to be adopted.

> So far as the world is concerned, past ages are far from countless; the ages to come are numbered; no one age has resembled its predecessor, nor will any future time repeat the past. The estimates of geologists must yield before more accurate methods of computation, and these show that our world cannot have been habitable for more than an infinitely insufficient period for the execution of the Darwinian transmutation.[46]

Also like Kelvin, he chose Darwin's calculation of the denudation of the Weald as an example of the errors which could arise from an indiscriminate reliance upon the principle of uniformity. The erosion of the Weald, he said, might have taken place a thousand times faster or a million times slower than Darwin had estimated. The available data was simply too meagre for judgment. "The whole calculation," he concluded with biting precision, "savours a good deal of that known among engineers as guess at the half and multiply by two."[47]

Of the other critics who used Kelvin's chronology to refute Darwin, perhaps the most notable was St. George Mivart (1827-1900). By the time Mivart's *On the Genesis of Species* appeared in 1871 both Huxley and Wallace had yielded to Kelvin's arguments without abandoning Darwinism. But Mivart refused to accept the possibility of such a compromise. A significant part of his criticism of the *Origin* hinged upon the absence of any trace of the transitional forms that must have existed if one species evolved from another. This was an objection that Darwin had attempted to answer by emphasizing the imperfection of the geological record. If the time available were much reduced, however, Darwin's defense would be severely weakened, and this was the point Mivart hoped to make with the help of Kelvin's physics. He made no pretext of being able to evaluate the validity of the physicist's arguments, but argued that the fact that they had not been refuted spoke loudly in their favor. He also contended that since neither biology nor geology alone could resolve the question of how imperfect the geological record really was, the results of physics had to be considered. In Mivart's opinion there was no question but that natural selection had to give way, and to that end he strongly endorsed Kelvin's conclusions.[48]

Darwin was greatly troubled by the use of Kelvin's arguments, and in his letters after 1868 he frequently alluded to his doubts and uncertainties. Early in 1869, for example, he complained to Wallace that "Thomson's views on the recent age of the world have been for some time one of my sorest troubles."[49] He had reason to be troubled, for Kelvin's most sustained attack upon natural selection and the uniformitarian view of time came between 1868 and 1871, just when the last two editions of the *Origin* were being prepared. And neither Darwin nor his supporters could find adequate refutations for his arguments.[50] The only ray of hope seemed to come from Croll, and Darwin grasped it eagerly.

Like Lyell, Darwin was first attracted by Croll's glacial hypothesis. For years he had hoped to find evidence that the ice ages had not occurred simultaneously throughout the world, but until he read Croll had begun to doubt that the hope would ever be realized.[51] He consequently greeted Croll's work enthusiastically, even the parts dealing with geological time. As he wrote in September 1868:

> I have never, I think, in my life been so deeply interested by any geological discussion. I now first begin to see what a million means, and I feel quite ashamed of myself at the silly way in which I have spoken of millions of years.[52]

As he prepared the fifth edition of the *Origin*, Darwin asked for further elaboration on the question of time, and was greatly pleased to learn of Croll's efforts to extend the physical estimate of the sun's age.[53] Nonetheless he remained disturbed by the fear that even Croll's most generous estimate of solar heat would still allow too little time. Thus in January 1869 he wrote again:

> Notwithstanding your excellent remarks on the work which can be effected within a million years, I am greatly troubled at the short duration of the world according to Sir W. Thomson, for I require for my theoretical views a very long period before the Cambrian formation.[54]

There appeared to be no way out. Whatever time might be added to the age of accumulated strata had to be bought at the expense of the Precambrian. The sun limited the age of both, and natural selection demanded ample time in both. The fifth edition of the *Origin* bears witness to Darwin's discomfort and his reluctant attempts to compromise. The Lamarckian reliance upon the direct effects of environment and the use and disuse of parts, never entirely absent from the *Origin*, became more pronounced as Darwin attempted to speed up the process of evolution. And where the denudation of the Weald had once illustrated the immensity of time, he substituted a detailed discussion of Croll's views on the inability of man to conceive of millions of years and on the consequent errors of geologists in attempting to express geological time in years. He also endorsed Croll's assessment of how rapidly denudation actually takes place, and though clearly still hoping that the 60 million year estimate of time since the beginning of the Cambrian might be extended, conceded that great changes could have taken place during the time allotted.[55]

Compromise was not total retreat, and there were occasions when Darwin felt a surge of optimism. One such occasion came early in 1869 when he read an anonymous review of Kelvin's opinions and of the recent controversy between Kelvin and Huxley. The reviewer, who turned out to be P. G. Tait, Kelvin's long-term collaborator, was obviously a physicist and obviously strongly partisan in Kelvin's favor. But paradoxically, he took an excessively hard line in insisiting that Kelvin had been over generous in his estimate of the sun's age. The true limit of time, he asserted, must be closer to 10 million years than to Kelvin's 100 million. Darwin saw a faint ray of hope, however, in the fact that the physicists disagreed among themselves. As he gleefully remarked to Hooker, geologists should be well warned and be careful in accepting the physicists' views. Physics, after all, had been wrong before. "Nevertheless," he continued, "all uniformitarians had better at once cry 'peccavi'—not but what I feel a conviction that the world will be found rather older than Thomson makes it, and far older than the reviewer makes it. I am glad I have faced and admitted the difficulty in the last [fifth] edition of the *Origin* . . . . "[56]

Darwin's concern over the limitation of the earth's age was not shared by his co-discoverer of natural selection, Alfred Russel Wallace (1823-1913). Although initially cautious about accepting Kelvin's results, the arguments of Geikie and Croll had quickly convinced Wallace that the uniformitarians had greatly overestimated the time necessary for the actions of denudation and sedimentation.[57] By late 1869 he was willing to accept Kelvin's arguments, and before the end of the year enthusiastically wrote Darwin that he had found a way out of the dilemma of time.[58] His solution appeared in a fascinating short paper early the following year.[59]

The basis for Wallace's new hypothesis was Croll's theory of the cause of the ice ages. A year earlier he had still leaned toward Lyell's geological explanation of widespread glaciation and had given Croll's astronomical hypothesis only secondary consideration, but thanks to Darwin's enthusiasm he completely reversed his views.[60] Not only did he accept Croll's hypothesis as the best key to determining geological time, he also saw in it a previously overlooked factor which he believed must influence the operation of natural selection. Referring to Croll's calculation of the changes in the earth's orbital eccentricity during the last three million years Wallace observed that during most of that time the eccentricity was considerably higher than it is at present. In fact, for the last 60,000 years the earth has been in a period of uncommonly low orbital eccentricity. According to Croll's hypothesis, therefore, the earth must now be experiencing an abnormally prolonged period of climatic stability. Applying this conclusion to the problem of species change through natural selection, Wallace drew an amazing yet logical conclusion.

Croll's hypothesis asserted that during periods of high orbital eccentricity radical changes in climate would occur every 10,500 years (half the period of a complete precession) as glaciers covered first one hemisphere and then the other. Wallace argued that these changes would be particularly pronounced in the extratropical regions where they would result in vast migrations of both plants and animals. This almost constant migration, he continued, would intensify the competition between allied species and force many to extinction. And finally, in a notable example of continued Lamarckian influence, he suggested that these "altered physical conditions would induce variation" among the individual organisms and thus accelerate the process of change. All in all, he concluded:

> . . . we should have all the elements for natural selec-
> tion and the struggle for life, to work upon and de-
> velop new races. High eccentricity would therefore
> lead to a rapid change of species, low eccentricity to a
> persistence of the same forms; and as we are now, and

have been for 60,000 years, in a period of low eccen-
tricty, *the rate of change of species during that time may be no
measure of the rate that has generally obtained in past geological epochs.*[61]

Turning next to the problem of the earth's age itself, Wallace again
looked to the arguments of Croll and Lyell. He had already been
convinced that Lyell had greatly underestimated the rates of geological
activity, and thus not surprisingly, he favored Croll's estimate of 80,000
years as the time since the last ice age rather than Lyell's choice of
nearly one million years. Nonetheless, he believed that Lyell's attempt
to measure the duration of geological epochs from the changes in the
species of marine mollusca would still prove valuable. Lyell had not
only chosen the earlier period of eccentricity for the glacial epoch,
however, he had also overlooked the 60,000 years of relative stability
immediately preceding the present age. Taking all of these factors into
account Wallace asserted that 100,000 years rather than one million
years would probably be a better (although perhaps still excessive)
estimate for time required for the extinction of 5% of the existing
species of marine mollusca. This substitution would reduce Lyell's
results by a factor of ten and would fix the age of the Cambrian
formation at only 24 million years. Thus, Wallace's hypothesis could
easily accommodate Darwin's demands for a long Precambrian period
for the development of life and still fit within Kelvin's 100 million years
for the total age of the earth. The Precambrian would in fact be three
times longer than all post-Cambrian time; and the rate of biological
change during that and subsequent periods would on the whole be
faster than the relatively slow change in present species.

Darwin was not convinced. Returning a draft of the article he commented: "your argument would be somewhat strengthened about organic changes having been more rapid, if Sir W. Thomson is correct
that physical changes were more violent and abrupt." He was not yet
ready to make such a concession. The real rub, however, was simply
that: "I have not yet been able to digest the fundamental notion of the
shortened age of the sun and earth."[62]

Despite mounting pressures from the seemingly irrefutable arguments of the physicists, Darwin never really accepted Wallace's compromise. He neglected it even in his letters to Wallace, as for example,
in 1871 when he was searching for a reply to Mivart: "I can say nothing
more about missing links than I have said. I should rely much on
pre-Silurian times; but then comes Sir W. Thomson like an odious

spectre."[63] Wallace's solution was clearly not the one he sought, and when he turned to the question of time in the sixth edition of the *Origin*, it was Croll's calculation that he discussed rather than Wallace's. There had been one significant effect, however, for after discussing both Kelvin and Croll and the difficulty in reconciling natural selection with even the greatest period that their combined hypotheses would allow for the Precambrian, he continued:

> It is, however, probable, as Sir William Thompson [sic] insists, that the world at a very early period was subjected to more rapid and violent changes in its physical conditions than those now occurring; and such changes would have tended to induce changes at a corresponding rate in the organisms which then existed.[64]

This was, of course, the very concession that he had refused to make to Wallace two years earlier.

Like all of Darwin's compromises, his concessions to Wallace and Kelvin included a hedge. Thus his final published statement on the subject of time attempted to retrieve with one hand what he felt obliged to concede with the other:

> With respect to the lapse of time not having been sufficient since our planet was consolidated for the assumed amount of organic change, and this objection, as urged by Sir William Thomson, is probably one of the gravest as yet advanced, I can only say, firstly that we do not know at what rate species change as measured in years, and secondly that many philosophers are not as yet willing to admit that we know enough of the constitution of the universe and of the interior of our globe to speculate with safety on its past duration.[65]

In effect, Darwin asked for a suspension of judgment because he still could not accept the idea that Kelvin's allowance of time would be adequate for what he saw as the necessarily slow progress of Precambrian evolution. But he stood against the tide. Geikie, Croll, and Wallace were heralds of the changed geological opinions that were to dominate the next few decades, and even Huxley, rising to defend geology and natural selection from Kelvin's attack, found that he could do little more than lead a graceful retreat.

The Kelvin-Huxley Debate

That Thomas Henry Huxley (1825-1895) should respond to Kelvin's challenge to the uniformitarian and Darwinian theories was almost inevitable. He was both Darwin's self-appointed "bulldog" and the president of the Geological Society of London at the time when Kelvin's influence first came to be strongly felt; and he was certainly not adverse to a good fight. He therefore took the occasion of his presidential address in 1869 to reply to Kelvin's assertion of a year earlier that "a great reform in geological speculation seems now to have become necessary."[66] Kelvin's rebuttal, delivered like his original challenge before the Geological Society of Glasgow, came within the month, and there, uncharacteristically for both, the matter was allowed to rest. Neither Huxley nor Kelvin had attended the other's lectures, and neither was primarily a geologist; but the nature of their exchange left no doubt that a debate had in fact taken place, and its effects influenced geological theory for decades.

When Kelvin initiated the debate in February 1868, the revisions for the fifth edition of the *Origin* were yet to be begun, Geikie's paper on denudation was yet to be delivered, and the full account of Croll's theory was still at the printers. And though Phillips, Croll, and a few others had publicly endorsed the physical limits on the earth's age, the recent edition of Lyell's *Principles* had again raised the spectre of the earth as a great perpetual motion machine. Kelvin consequently felt, with some justification, that his arguments had been ignored and that geologists in general were still neglecting the basic principles of natural philosophy.

Nonetheless, the attack began mildly. At the end of the eighteenth century, he conceded, geology clearly needed more time than the dogmatic diluvian hypotheses of the period allowed, and the geologists had justifiably sought to break free from the restrictions of Mosaic chronologies. He also praised the efforts made since then to make geology into a truly exact science. Some modern geologists, however, had gone too far in their reaction against the old hypotheses, and the time had long passed for them to reassess the principle of uniformity and its demands for unlimited time. Curiously, after his ringing call for reform, Kelvin chose to resurrect John Playfair, the spokesman for a much earlier generation, as the focus for attack. Yet the implication was plain; geology had not kept up with the advances in physical science.

Kelvin's criticism centered on the accusation that Playfair had confused *present order* or *present system* with the *laws now existing*. The laws of nature had not changed, Kelvin agreed, but the condition of the earth had changed continuously. Indeed, it was just because the second law of thermodynamics was unvarying, that one could be certain that the energy available for geological activity must have decreased constantly since the world began. Friction alone must have cost the earth a measurable part of its store of energy; and Kelvin used Playfair's neglect of this point, and the neglect of others including the mathematicians Laplace and Lagrange, as a springboard for proposing his theory of tidal retardation. The calculation of time which this argument permitted was but a small part of his concern, however. His other arguments had already provided much more explicit results than it could supply. His real aim was to show, once and for all, that the laws of thermodynamics utterly negated the possibility that either continuously uniform or perfectly cyclic processes could exist in a finite world, and therefore that geological uniformity could not be a law of nature. Kelvin was convinced, moreover, that so long as geologists continued to cling to uniformitarianism, geology would never take its place as a truly exact science. Thus it was his contemporaries rather than Playfair or Hutton whom he challenged with the words: "It is quite certain that a great mistake has been made—that British popular geology at present is in direct opposition to the principles of natural philosophy."[67]

Huxley, a formidable champion in the arena of public debate, rose to answer the challenge. In this instance, however, his efforts must be praised more for their vigor than their strength. His address was characteristically lucid and elegant, but it seldom came to grips with the fundamentals of Kelvin's argument. To use his own metaphor, he sought by "mother-wit" to act as an attorney for the defense of geology.[68] But though the jury before whom he tried his case—the Geological Society of London—was friendly, mother-wit alone could not combat the carefully reasoned, physically and mathematically correct arguments of his opponent.

Kelvin had attacked British "popular geology" which to him meant uniformitarianism, but in choosing to focus his attack on Playfair rather than on Lyell or another contemporary geologist, he provided an excellent opening for rebuttal. Huxley seized on it immediately. If Playfair and Hutton had improperly assumed that the intensity of the earth's activity has never changed, then point out the fallacy, he challenged, but do not assume that a valid attack upon them is also a valid attack upon modern geology. In effect, he accused Kelvin of attacking a straw man as he continued:

> I do not suppose that, at the present day any
> geologists would be found to maintain absolute Un-
> iformitarianism, to deny that the rapidity of the rota-
> tion of earth *may* be diminishing, that the sun *may* be
> waxing dim, or that the earth itself *may* be cooling.
> Most of us, I suspect, are Gallios, "who care for none of
> these things," being of the opinion that, true or ficti-
> tious, they have made no practical difference to the
> earth, during the period of which a record is preserved
> in stratified deposits.[69]

Certainly Huxley's assertion here was not entirely unjustified, but
neither was it particularly accurate. In the year since Kelvin's challenge
had been issued both Geikie and Croll had come out strongly in
support of the belief that geological theory could be reconciled with his
limited time scale. A few other geologists had also abandoned strict
Lyellian uniformity, and some indeed had never really accepted it. But
the majority of their fellows seldom confronted the question directly.
They were content to assume uniformity for the sake of description
and explanation, and to accept the idea that geological time was virtually limitless if not actually infinite. Huxley's remark had thus sac-
rificed accuracy for rhetoric and yet had not avoided conceding the
substance of Kelvin's argument.

The weakness of Huxley's position was still more apparent in his
treatment of Kelvin's quantitative estimates of time. He began by
asking whether it had ever been denied that 100 million years *may* have
been long enough for the purposes of geology; a question that could
have been answered emphatically yes had he chosen to answer. But
instead he reversed his ground completely to assert that two, three, or
four hundred million years made "all the difference" in Kelvin's
theory. He had missed the point entirely—or had carefully avoided
it—for Kelvin's position had been that if any limit whatever could be
placed upon the earth's age, it would refute uniformitarianism. It was
sheer sophistry, moreover, to criticize Kelvin's broad allowances for
error as vagueness, while in the same breath asserting that a change of a
factor of two or three in his results would negate his argument. Huxley
had sidestepped the fundamentals of Kelvin's objections to uniformitarianism, and yet in conceding the possiblility of limiting the earth's
age, he essentially admitted his willingness to abandon the indefinite
time of Lyell. Darwin too must have felt abandoned when he heard:

Biology takes her time from geology. The only reason we have for believing in the slow rate of the change in living forms is the fact that they persist through a series of deposits which, geology informs us, have taken a long while to make. If the geological clock is wrong, all the naturalist will have to do is to modify his notions of the rapidity of change accordingly.[70]

Certainly Darwin could take little comfort in these remarks since Fleeming Jenkin had already shown that natural selection alone could not adequately account for species change within Kelvin's limited time.

Huxley's address was not all mere sound and fury, however. Even his disregard for time was based on his belief that geology could and should accommodate the laws of physics. Thus the most significant part of his address concerned a proposed reform in geological theory that, while not exactly what Kelvin had in mind, would alleviate some of the major difficulties. Kelvin had attacked only uniformitarianism as British "popular geology," and Huxley questioned whether this was either just or accurate since catastrophism still claimed many able adherents. Reform might be necessary, but Huxley asserted that it should involve eliminating the weaknesses of both uniformitarianism and catastrophism while preserving the strengths of both. After all, he argued, there was actually no basic antagonism between catastrophism's practically unlimited bank of force and uniformitarianism's practically unlimited bank of time. "On the contrary, it is very conceivable that catastrophes may be part and parcel of uniformity."[71] The real disagreement, Huxley believed, arose from the antiprogressional bias of the uniformitarians, and from the catastrophists' neglect of the present as the only reliable guide to the past. Both schools had failed to approach geology in what he termed the aetiological sense; that is, neither had attempted "to deduce the history of the world as a whole from both the known properties of the matter of the earth and the conditions in which the earth has been placed."[72] Huxley's new approach, which he dubbed Evolutionism, would address itself to just this problem. It would allow for geological forces in the past that were of greater power or of a different nature than those in the present—in other words, it would allow for catastrophes in a natural but not a supernatural sense—but would stress the necessity of studying present geological processes as the guide to understanding the record of the past. Furthermore, by taking progression rather than uniformity as the key to the earth's history, evolutionism would recognize that geological change was a function of time. It would if necessary accommodate even the limited time scale of Kelvin. Clearly, Huxley's evolutionism bore a marked resemblance to the kind of uniformitarianism proposed by Geikie, and under either name it had a profound effect.

A final point in the address raised the question of mathematical certainty, and here he confronted Kelvin most directly. In looking at the physical arguments, Huxley granted Kelvin the accuracy of his mathematics and the probable validity of his principles, but not the necessary validity of his assumptions. Repeatedly he pointed out how Kelvin had required his readers to suppose this or assume that, while he stressed the hypothetical nature of these assumptions and argued that other suppositions had an equal right to consideration. Singling out Kelvin's most recent argument, the case of the tidal retardation of the earth, Huxley assailed it as an example of how "the admitted accuracy of mathematical process" had been allowed "to throw a wholly inadmissible appearance of authority over the results obtained by them." Kelvin's mathematics were no doubt beyond question, but:

> Mathematics may be compared to a mill of exquisite
> workmanship, which grinds you stuff of any degree of
> fineness; but, nevertheless, what you get out depends
> upon what you put in; and as the grandest mill in the
> world will not extract wheat-flour from peascod, so
> pages of formulae will not get a definite result out of
> loose data.[73]

Huxley's analogy was frequently quoted, but the full weight of his cautionary remarks was not immediately appreciated. Although in retrospect it seems his strongest argument by far, Kelvin's opponents failed to capitalize on it until late in the century. Thus for nearly two decades the most influential part of the address was not Huxley's warning about the uncertainties of Kelvin's assumptions, but his assertion that biology could take its time from geology.

Kelvin responded eagerly. His rebuttal, delivered only two weeks later, was obviously hurried, but it contained a point by point refutation of Huxely's position.[74] Lacking the gift of prophecy, Kelvin could not appreciate the value of his opponent's *evolutionism*. He dismissed it facilely, but with some justice, as essentially the geology that he had learned at Glasgow University thirty years before. Kelvin's dispute, after all, was with uniformitarianism, and he emphatically denied Huxley's contention that few geologists were true uniformitarians. He also vigorously condemned the flippancy of Huxley's appeal to mother wit and the know-nothing attitude implied in the assertion that geology cared nothing about whether the sun were cooling or the earth's rotation slowing down. This was precisely the attitude that he believed had led geology into conflict with natural philosophy and that he sought to combat. As for the assertion that the limitation in geological time made no difference to biology, he replied:

> The limitations of geological periods, imposed by physical science, cannot, of course, disprove the hypothesis of transmutation of species; but it does seem sufficient to disprove the doctrine that transmutation has taken place through "descent with modification by natural selection."[75]

Darwin himself could only reluctantly concur in this judgment.

Kelvin's address contained nothing original. In fact, it consisted largely of extended quotations to illustrate his points, but the quotations were effectively used. He turned Huxley's own words upon him, bombarded his listeners with quotations from geologists proclaiming the virtual limitlessness of the earth's age, and cited Geikie and Phillips to show how the geological evidence could agree with natural philosophy when "properly interpreted." With regard to Huxley's more substantial argument, Kelvin admitted that his calculations involved many assumptions, but vehemently denied the accusation that his mathematics gave too much authority to results from "loose" data. His position remained firm that the wide limits which he had allowed in his results were quite sufficient to account for any errors in either his data or assumptions. One hundred million years was the maximum probable duration of life on earth that his calculations would allow, but he remained convinced that even 400 million years would not satisfy the demands of the uniformitarians.

Although the exchange between Huxley and Kelvin contributed nothing new to the solution of the problem of the earth's age, it inaugurated a new era in geological speculation that went far beyond the single question of time. Huxley's defense of uniformitarianism had failed to materialize. Instead he had championed a new approach to the problems of physical geology, an approach which could be made compatible with Kelvin's conclusions about the age of the earth and which was already being followed by a few geologists like Geikie and Croll. The debate succeeded, however, in bringing the question of geological time into the public forum and in generating considerable public interest. No longer could Kelvin complain that his arguments were ignored by the geologists. The apparent strength of his position had been demonstrated, and the weight of both scientific and public opinion had begun to swing decidedly in his direction.

REFERENCES

[1]Geikie, A. (1899), *Presidential Address*, p. 198.

[2]Hopkins (1839), *Precession and Nutation*.

[3]Examples of those geologists supporting Hopkins and Kelvin on the solidity of the earth include: Fisher (1868b), *Elevation of Mountain Chains;* Hunt (1867), *Chemistry of the Primeval Earth*; Scrope (1868), *The Supposed Internal Fluidity of the Earth*; and Shaler (1868), *Formation of Mountain Chains*. Those opposed included: De Launay (1868), *Internal Fluidity of the Terrestrial Globe;* Forbes (1867), *Chemistry of the Primeval Earth;* Phillips (1869), *Vesuvius*, p. 329; and Ward (1868), *Internal Fluidity of the Earth*.

[4]Kelvin's address to the B. A. in 1861, the forerunner of "On the Age of the Sun's Heat," was mentioned only by title in *The Geologist*, 1861, 4:547, while several other papers were reprinted entirely or in detailed abstracts. The two landmark papers of 1862 went unnoted in any major geological publication.

[5]Kelvin (1866a), *Doctrine of Uniformity*, pp. 6-9.

[6]Phillips (1837), *Treatise on Geology*, I:4-18. Quotation, p. 18.

[7]Phillips (1860b), *Life on Earth*, pp. 122-137.

[8]*Ibid.*, pp. 126-127.

[9]See Phillips' Letter to Kelvin, 12 June 1861. Thompson (1910), *Kelvin*, I:539. Quoted in part in Chapter 2.

[10]Phillips was in fact responsible for several of the first public statements recognizing Kelvin's work. (See: Phillips (1864), *Address to the Section of Geology*, pp. 175-180; and Phillips (1865), *Presidential Address*, pp. liii-liv.)

[11]Geikie, A. (1867), *Geological Time*.

[12]Geikie, A. (1871), *Modern Denudation*.

[13]*Ibid.*, pp. 188-189.

[14]Croll (1864), *Change of Climate*, p. 137.

[15]Irons (1896), *Life of Croll*, pp. 34, 113-114, 165-169.

[16]*Ibid.*, pp. 32-35, 61, 154-155. (Also see: Croll (1868), *On Geological Time, 36*:143; and Croll (1871a), *Determining the Mean Thickness of Sedementary Rocks*, p. 100.)

[17]Croll (1868), *On Geological Time, 35*:374-375. Also repeated in Croll (1875b), *Climate and Time*, pp. 326-327. The book *Climate and Time*, published in 1875, contained the substance of Croll's numerous articles on geochronology and glacial theory which were published between 1864 and 1875. Croll's ideas were already well known among scientists before the book appeared, but it gave them much wider currency.

[18]Croll (1871c), *Age of the Earth.*

[19]Croll (1868), *On Geological Time,* 35:368-74.

[20]Herschel (1830), *Preliminary Discourse,* pp. 145-47; Herschel (1835), *Astronomical Causes;* Herschel (1849), *Outlines of Astronomy,* pp. 215-219; Arago (1834), *State of the Terrestrial Globe;* Adhémar (1860), *Révolutions de la Mer,* pp. 37-41, 80-85, 338-353 (I have been unable to consult the 1st edition of Adhémar's work which appeared in 1842); Humboldt (1848), *Cosmos,* IV:458-459.

[21]Croll (1868), *On Geological Time, 35*:363-368, *36*:146-154.

[22]Croll (1864), *Change of Climate;* Croll (1866a), *Eccentricity of the Earth's Orbit;* Croll (1866b), *Submergence and Emergence of Land;* Croll (1867a), *Excentricity of the Earth's Orbit;* and Croll (1867b), *Obliquity of the Ecliptic.*

[23]Croll (1867a), *Excentricity of the Earth's Orbit.*

[24]Croll (1868), *On Geological Time, 35*:366-368.

[25]Walter Cannon has shown that in spite of Lyell's tremendous influence, catastrophism remained an important geological doctrine through the early 1850s. (See: Cannon (1960), *Uniformitarian-Catastrophist Debate.*) But it is nonetheless clear that uniformitarianism had risen steadily in importance and by the 1860s had emerged as the dominant, if not altogether unchallenged, view of British geologists.

[26]Lyell, K. (1881), *Life of Lyell,* I:114-117 (Letter from Lyell to J. Croll 13 Feb. 1865) and I:406-407 (Letter from Lyell to Prof. Heer 21 Jan. 1866).

[27]Lyell, C. (1867-68), *Principles,* 10th ed., I:271-282, 291-301.

[28]Lyell, C. (1877), *Principles,* 11th ed., I:284-296.

[29]Lyell, C. (1850b), *Principles,* 8th ed., pp. 512-525.

[30]Lyell, C. (1867-68), *Principles,* 10th ed., II:212-213.

[31]*Ibid,* II:230-232.

[32]Darwin, C. (1859), *Origin,* 1st ed., p. 282.

[33]*Ibid.,* pp. 285-287. This estimate is excessive even by today's standards which put the origin of the Weald at about 135 million years.

[34]Anonymous (1859), *Darwin's Origin.*

[35]Phillips (1860b), *Presidential Address,* pp. lii-lv; Phillips (1860b), *Life on Eath,* p. 130. Phillips had crossed swords with Lyell and Darwin as early as 1838, shortly after the latter's return on the *Beagle.* (See: Lyell, K. (1881) *Life of Lyell,* II:39-41. This letter also appears misdated as 1858, *Ibid.,* II:281-282.) It is apparent from Darwin's correspondence with Lyell in 1859-60 that he regarded Phillips' views with reluctant respect. (See: Darwin, F. (1888), *Life of Darwin,* II:309, 349; Darwin, F. and Seward (1903), *More Letters of Darwin,* I:127, 130, 141.)

[36]Darwin, C. (1959), *Origin, Variorum,* p. 484.

[37]Darwin, F. (1888), *Life of Darwin,* II:264.

[38]*Ibid.,* II:350. This statement was made before Darwin read *Life on Earth* (See: *Ibid.,* II:349) which he subsequently criticized quite harshly. (See: Letters to J. D. Hooker, 15

Jan. 1861, and A. Gray, 5 June 1861, *Ibid.*, II:358, 373-374.) Thus, although he left the calculation out of the third edition, he did not include the proposed footnote.

[39]Darwin, F. and Seward (1903), *More Letters of Darwin*, II:139.

[40]Kelvin (1871c), *Presidential Address*, pp. 197-205. Huxley referred to the theory as "Thomson's 'creation by cockshy'—God Almighty sitting like an idle boy at the seaside and shying aerolites (with germs), mostly missing, but sometimes hitting a planet!" (See: Huxley, L. (1918), *Life of Hooker*, II:126.)

[41]Kelvin (1889), *On the Sun's Heat*, p. 422.

[42]Jenkin (1867), *The Origin of Species*.

[43]Thompson (1910), *Kelvin*, I:408-409, 552-553.

[44]Stevenson (1887), *Memoir of Jenkin*, I:lxi.

[45]Jenkin (1867), *The Origin of Species*, pp. 294-305.

[46]*Ibid.*, p. 295.

[47]*Loc cit.*

[48]Mivart (1871), *Genesis of Species*, pp. 142-57.

[49]Marchant (1916), *Letters of Wallace*, I:242. Letter dated 14 April 1869.

[50]For some indication of the magnitude of Kelvin's influence at this time see: Ellegârd (1958), *Darwin and the General Reader*, pp. 237-38.

[51]Darwin, F. and Seward (1903), *More Letters of Darwin*, I:460-465. Letters to Lyell, Hooker, Huxley and H. W. Bates between 1 Nov. 1860 and 26 Mar. 1861.

[52]*Ibid.*, II:211. Letter dated 19 Sept. 1868.

[53]Irons (1896), *Life of Croll*, pp. 200-203, 216-221.

[54]Darwin, F. and Seward (1903), *More Letters of Darwin*, II:163. Letter dated 31 Jan. 1869.

[55]Darwin C. (1959), *Origin, Variorum*, pp. 482-486.

[56]Darwin, F. and Seward (1903), *More Letters of Darwin*, I:313-314. Letter dated 24 July 1869.

[57]Wallace (1869a), *Geological Climates*, pp. 375-376.

[58]Marchant (1916), *Letters of Wallace*, I:246. Letter dated 4 Dec. 1869.

[59]Wallace (1870), *Geological Time*.

[60]Wallace sent a prepublication copy of "Geological Climates," which was a review of the new editions of Lyell's *Principles and Elements*, to Darwin early in 1869. Darwin immediately responded with the opinion that Wallace had not placed as much confidence in Croll's work as he, Darwin, was willing to do. (See: Marchant (1919), *Letters of Wallace*, I:242. Letter dated 14 April 1869.) There can be little doubt that this opinion was in some degree influential in Wallace's subsequent about-face and his adoption of Croll's hypothesis in "Geological Time."

[61]Wallace (1870), *Geological Time*, p. 454. (His italics)

[62] Marchant (1916), *Letters of Wallace,* I:250-251. Letter dated 26 Jan. 1870. Actually, Darwin was confusing Wallace's argument for rapid climatic change due to alterations of astronomical conditions with Kelvin's argument for greater past meteorological and plutonic activity due to higher temperatures in the earth and sun. The two arguments have little in common, but I have found no record of Wallace correcting Darwin.

[63] *Ibid.,* I:268. Letter dated 12 July 1871.

[64] Darwin, C. (1959), *Origin, Variorum,* p. 513.

[65] *Ibid.,* p. 728.

[66] Kelvin (1871a), *Geological Time,* p. 10.

[67] *Ibid.,* p. 44.

[68] Huxley, T. H. (1869), *Geological Reform.*

[69] *Ibid.,* p. 329. Huxley's allusion was to the Bible, Acts 18:17. Gallio, the Proconsul of Achaea, refused to try Paul under Roman law for breaking Jewish law. Jewish law *was not his province* and he "cared for none of those things." Huxley's reference was perhaps more pointed than it appears at first.

[70] *Ibid.,* p. 331.

[71] *Ibid.,* p. 327.

[72] *Ibid.,* p. 322.

[73] *Ibid.,* p. 335-336.

[74] Kelvin (1871b), *Geological Dynamics.*

[75] *Ibid.,* pp. 89-90.

IV

The Triumph of
Limited Time

Geological theory was clearly in a state of transition by the time of the exchange between Kelvin and Huxley. Indeed, it had been in continuous transition since the appearance of Lyell's *Principles of Geology* in 1830. More than anyone else, Lyell had elevated time to a position of primary importance in the explanation of geological dynamics. He had substituted the immensity of time for the immensity of force as the key which would unlock the secrets of the earth's changing surface, and for thirty years this view steadily gained adherents among British geologists. By 1859 it was clearly the dominant geological theory in England. Ten years later that position just as clearly was being modified if not altogether reversed. In 1859 Darwin attacked the uniformitarian assumption of non-progressionism in biology, and in 1862 Kelvin attacked the same assumption in geology. Their combined effect was such that by the end of the decade a strictly Lyellian view was more and more frequently conceded to be untenable.

The critical element in the new geology was directional change. To be sure, neither Lyell nor his followers had denied the reality of geological changes. The importance of minute changes reinforced over vast times was as much a central premise of uniformitarianism as it was of evolutionary biology or the new evolutionary geology. But the changes which Lyell visualized were local; they oscillated between rather narrow limits; and in the end, they resulted in an essentially unchanging, dynamically balanced, steady-state earth.[1] Such a conception was totally incompatible with the directionalism which Darwin had reintroduced into biology and which Kelvin, through the laws of thermodynamics, had revived in geology. Kelvin and Darwin believed their views to be mutually exclusive, however, and thus it was Huxley, unable either to reject Darwin or to refute Kelvin, who pronounced the two compatible.

Geology meanwhile reeled under the double blow, and the ambivalence which the new conceptions created in the minds of many geologists was reflected in the reviews of the Kelvin-Huxley debate. Much of the reaction was understandably partisan. Darwinians rallied immediately behind Huxley, but often failed to recognize the essential compromise he had proposed. Kelvin's defenders gathered more slowly, but their numbers increased steadily. One thing at least was certain: geologists could no longer afford to ignore Kelvin's ideas. For the first time, his arguments were seriously discussed in the leading geological journals and, perhaps equally important, in popular non-scientific journals and reviews.[2] The more perceptive of the commentators recognized the importance of the introduction of direction into geology and gave only secondary attention to Kelvin's numerical results. The more partisan reviewers, on the other hand, allowed the limitation of geological time to eclipse all other considerations. Both groups were united, however, in calling into question certain of the fundamental assumptions of Lyell's uniformitarianism. In the end, time was accepted as finite. A belief in the uniform action of geological processes in *kind* if not in *degree* survived, but the idea of a steady-state earth gave way to the directionalism of Huxley's evolutionism.

As the preceding chapter showed, the reversal in geological thought began well before the Kelvin-Huxley debate. Despite Kelvin's assertions to the contrary, no clear cut consensus of "British popular geology" existed in 1869, if indeed it had ever existed. Nonetheless, the debate and its aftermath did serve to give divergent views greater prominence, and thus can appropriately be seen as a kind of watershed. Before 1869, for example, the popular and influential journal, *Geological Magazine*, had given scant notice to Kelvin's arguments. But it carried detailed reviews of both Huxley's address and Kelvin's rejoinder, and thereafter discussions of the earth's age from both the geological and the physical points of view appeared frequently in its pages. The reviews of the debate were themselves significant, for although neither supported Kelvin, only one defended uniformitarianism. The other warmly endorsed Huxley's evolutionary geology.

Huxley's reviewer praised the style, eloquence, and logic of his address, but disagreed sharply on the fundamental question of time.[3] He accused Huxley of conceding to Kelvin's limitations when no concession was necessary. Geology alone, the reviewer declared, indicates that the earth is vastly old, and unless the physicists could claim omniscience in accounting for every possible source of energy, he could see no grounds for limiting its age. He also reaffirmed his belief in the steady-state, anti-evolutionary view of the earth's history which had characterized Lyell's original hypothesis and condemned evolutionism as an attempt to reintroduce into geology the cosmological speculations that Hutton and Lyell had so carefully and justifiably thrown out. The reviewer thus provides a nearly perfect example of the kind of geological thinking Kelvin had attacked. But he was already an anachronism, abandoned even by Lyell whose earlier views he was defending.

Quite a different point of view was expressed in the review of Kelvin's "Geological Dynamics."[4] Although the article was presented as a synopsis of Kelvin's address, it actually gave considerable emphasis and support to Huxley's position. By emphasizing the views of Huxley and Phillips, the reviewer made clear his opinion that Kelvin's limitation of time was of little consequence to geology. He also sloughed off Kelvin's detailed response to Huxley by blandly asserting that the latter's objections remained practically unanswered. Perhaps the most significant part of the review, however, was the author's enthusiastic support of evolutionism, and his prediction that it would eventually absorb both the uniformitarians and the catastrophists. In this remark he was not only prophetic, but like Huxley, he had implicitly admitted the possibility of an entente between Kelvin and the geologists.

The popular press was generally less reserved in assessing the debate than the specialist journals and for that reason more exciting. An article in the *Pall Mall Gazette*, for example, would have done justice to a youthful Henry Brougham except for the praise it heaped upon Huxley.[5] In espousing Huxley's contention that there was no necessary disagreement between catastrophism, uniformitarianism, and evolutionism, the *Pall Mall* reviewer did not argue merely that Kelvin had been refuted but that he had no point to refute. He conceded that the bank of time allowed by Kelvin's 100 million years would be quite adequate for geology if it were necessary to accept it, but he doubted such a necessity. Borrowing Huxley's metaphor he asserted that "Sir W. Thomson puts uncertainty in his mill and expects it to come out as absolutely true." For the reviewer, the uncertainties were sufficient to destroy the value of Kelvin's evidence completely. Thus he confidently concluded: "We entirely agree with Sir W. Thomson that 'it is quite certain that a great mistake has been made,' but it is one similar in kind to Sancho Panza's attack on the windmill, and it has not been made by British popular geologists."[6]

An equally partisan but more detailed defense of Kelvin appeared a

few months later in the *North British Review*.[7] Written anonymously by
Kelvin's friend P. G. Tait (1831-1901), the article was nominally a
review of several of Kelvin's papers, Huxley's address, Jenkin's review
of the *Origin*, and several related works, but it actually concentrated on
the Kelvin-Huxley debate. It was the most enthusiastic endorsement of
Kelvin's position that had yet appeared, or for that matter, that was
ever to appear. Tait's bias was evident in his dogmatic assertions that
the results of natural philosophy were unquestionable. "The fact is," he
asserted, "that although many scientific men (*in Britain*) may attempt to
ignore it, Mathematics is as essential an element of progress in every
real science as language itself." Mathematics, however, could only be of
value when a science had passed the "beetle-hunting" or "crab-
catching" stage to which Tait assigned the opinions of the majority of
geologists. In his opinion geologists were incapable of even arguing on
the same plane with physicists about the age of the earth. Therefore, he
proclaimed:

> Let us then hear no more nonsense about the interfer-
> ence of mathematicians in matters with which they
> have no concern; rather let them be lauded for con-
> descending from their proud preeminence to help out
> of a rut the too ponderous waggon of some scientific
> brother.[8]

Tait looked upon the methods of the geologists as totally inadequate
for solving the problem of the earth's age. At best geological methods
could give only an inferior limit to geological time when what was
needed was a superior limit. But more to the point, even the geologists'
inferior limit was greater than the laws of physics would permit. Turning the tables on Huxley, Tait sharply criticized the uncertainties in the
geological methods and the uniformitarians' confusion between the
permanence of natural laws and the permanence of present conditions. At the same time, however, he utterly denied any justification in
Huxley's warning about the uses, abuses, and limitations of the
mathematics in Kelvin's arguments. There was not the slightests possibility of error, he asserted, in extrapolating back to the beginnings of
the earth and sun on the basis of their thermal loss. Uniformitarianism
was the villain that had led geology astray, but mathematics and physics
could set it back on the right path if only geologists could be brought to
recognize their superiority.

As far as Kelvin's particular methods and results were concerned,
Tait admitted that each was subject to certain errors due to the lack of

experimental data. Nonetheless, he accepted the theories themselves as sound, and even with the deficiencies in the data, contended that they gave far more accurate and reliable results than the methods of geology. If Kelvin had erred at all, it was in giving the uniformitarians the best possible benefit of the doubt. Kelvin had in fact been far too generous in allowing 100 million years for the age of the earth. A closer look at the evidence, Tait believed, would indicate an age of only 10 million to 15 million years *or less*. Thus even if Huxley were correct in saying that uniformitarianism would be content with 100 million or 300 million years, he was sure that physics would not long permit even this.

If Tait was strident and uncritical in his championing of Kelvin, he was nonetheless perceptive in picking out the flaws in Huxley's argument. He was also more than justified in accusing the biologist of trying to mask the deficiencies of his position by playing with words and metaphors. What Tait refused to recognize, however, was the fact that Huxley himself had been aware that he could not debate Kelvin on his own terms and had laid the ground for a tactical retreat. Of course Huxley's compromise would still not accommodate an earth only 10 million years old, but few people ever really considered that to be necessary. In fact, by advocating such a radically truncated estimate of time, Tait succeeded in generating far more opposition than support.

Not all of the reviews were so stridently partisan, nor did the lack of partisanship necessarily entail intellectual spinelessness. The *Edinburgh Review*, for example, carried an article which was a model of balance between support for the new hypotheses and reasoned caution.[9] The author enthusiastically endorsed the directionalist thesis in Huxley's evolutionism, which he viewed as a necessary consequence of the principle of the dissipation of useful energy. He also championed Kelvin's hypothesis of the fiery origin of the earth and sun, and drew upon a wide range of evidence from geology, solar spectroscopy, and the chemical analysis of meteorites to support his belief. The reviewer looked upon both Kelvin and Huxley as advocates of an evolutionary system which opposed the non-directional uniformity of the "more popular school" of geology, and in this he gave them both his warm support. On the question of quantitative chronology, however, he struck a very different note. Not only did he reject the chronologies of both Kelvin and the geologists, he denied the validity of measurements in years or even millions of years as meaningful units of time. In the reviewer's opinion the relationships between the geological periods gave clear evidence of progress in the earth's history, and he accepted such evidence as significant and incontestable, but it provided absolutely no clue to the age of the earth in years. Indeed, the stratified rocks were "as likely to have been deposited in one million years as in five hundred million years."[10] Kelvin's results, on the other hand, were of just as little value because of the uncertainties in the physical data available. Thus, although willing to grant the justice of some of Kelvin's protests against the looseness with which geologists had spoken of geological time, the reviewer refused to accept either his estimate of the earth's age or Huxley's argument that the estimate was sufficient for geology.

The *Edinburgh* reviewer proved to be an exception, and once the radical implications of uniformitarianism had been overthrown, Kelvin's numerical results came to be stressed more and more. After all, the laws of thermodynamics upon which his assumptions were based were well established by 1870 and his mathematics were generally acknowledged to be beyond question. His numerical results alone seemed to be legitimate subjects of dispute as he himself showed by periodically changing the limits of time that his arguments would permit. Nonetheless, his original estimate of 100 million years for the earth's age exerted a profound influence as geologists began their own search for a quantitative chronology. Convinced of a certain validity in Kelvin's arguments the geologists' question was no longer whether the earth was exceedingly old or relatively young, but whether its assumed duration was sufficient to account for the appearance of present geological formations. Few geologists were willing to concede that Kelvin's later estimates met this criterion, but increasingly their evidence seemed to indicate that his original 100 million years might well be adequate for the purposes of geology.

Despite the seeming widespread interest, and the number of references to Kelvin's results were legion, very few investigators actually attempted to make independent calculations of the earth's age. Between 1870 and 1900 they numbered perhaps a score, divided more or less evenly between physics and geology. Nonetheless, they produced an amazing variety of methods and hypotheses and an even more amazing homogeneity of results. Kelvin was the acknowledged leader of the physicists, and most of their methods were either variations or extensions of his arguments. Each of them, too, shared with him the necessity of making certain primary assumptions and of estimating certain quantitative parameters. The geologists, on the other hand, introduced different methods and used different kinds of data, but because the rates of geological processes were very imperfectly known, they found that geology alone provided few guidelines for exact calculations. Without some reference point their data could give only relative ages, and thus, like the physicists, they were forced to estimate many of the parameters needed to quantify their results. It is hardly surprising, therefore, that each investigator, whether physicist or geologist, sought to make reasonable assumptions which were in accordance with the available evidence and which gave reasonable results. Nor is it surprising, given the persuasive arguments Kelvin had provided for his position, that the *reasonable* assumptions most frequently made were those which yielded results within the range of his 100 million years.[11]

Geology and the Measurement of Time

One of the most significant features of geology during the second half of the nineteenth century, and one that had a profound effect on the geologists' conception of time, was the concern with measurement. The call for better mensural techniques and more detailed quantitative data was a frequent refrain at geological meetings as geologists self-consciously sought to raise their discipline above the level of description and to emulate the more established *exact* sciences of physics and chemistry. T. Mellard Reade, for example, might have been echoing Tait when, in 1876, he declared that "to make Geology essentially a Science, the mathematical method must step in to measure, balance, and accurately estimate."[12] The same sentiment was no less strong a quarter century later when William J. Sollas proclaimed: "Our present watchword is Evolution. May our next be Measurement and Experiment, Experiment and Measurement."[13] No doubt some of the impetus for quantification came from the outside as men like Kelvin and Reade brought a knowledge of physics and engineering to bear on the problems of geology, but the geologists themselves were no less adamant in their desire to develop the techniques that would somehow make their science more exact.

For the most part the emphasis upon quantification did not entail mathematization in any rigorous sense nor the development of deductive mathematical hypotheses like those of physics. It was rather a Baconian drive to measure and categorize that resulted in the accumulation of vast quantities of data with few theoretical guidelines for their interpretation. In the case of the age of the earth, this meant that despite the growth of quantification, few geologists were equipped to evaluate Kelvin's arguments critically, even at the end of the century. And thus even though they were genuinely impressed by his physical reasoning, they were most directly influenced by his choice of data and his numerical results.

Another result of this lack of a theoretical framework was that geologists were far from united in their opinions on how their data related to geological time. Some of the older generation, including A. C. Ramsay (1814-1891), Joseph Prestwich (1812-1896), and James D. Dana (1813-1895), flatly rejected the idea that geological measurements could ever date the age of the earth in years.[14] Others, generally from a younger generation such as Geikie (1835-1924), Sollas (1849-1936), and Osmond Fisher (1817-1914), were willing to accept Kelvin's conclusions and adjust geological speculation accordingly.[15] But a small, vocal minority continued actively to develop quantitative methods with which to determine the earth's age from geological evidence alone. The three groups were united on one point, however: the limitless eons of the mid-century uniformitarians were no longer acceptable. The earth did have a finite and definable age.

The most common method of estimating the earth's age from geological evidence was to use the rate of either sedimentation or denudation as a timekeeper. Basically the method involved determining the total aggregate thickness of sedimentary rocks and the average rate at which they were built up by the deposition of sediment, and then taking the ratio of the two parameters as a measure of time. In practice, however, none of the parameters necessary for this calculation were known exactly. Some, such as the thickness of various stratigraphic layers, were subject to measurement; but the problems of combining data from widely scattered sources involved introducing assumptions that compromised the aim of precision. Other parameters, notably the rate of sedimentation, were beyond hope of direct measurement. The rate of sedimentation was directly related to the rate of denudation which could be measured, but what the quantitative ratio between the two actually was remained open to question. Certainly the area denuded by past erosion and the area concurrently receiving the sediment were beyond direct determination. The geologists were consequently forced to make the best estimates possible consistent with the data available, but the range of possible variations within the same basic method was enormous, and the preconceptions guiding each individual investigator played a decisive role in the results obtained.

Two variations on the use of denudation as a timekeeper have
already been discussed, Charles Darwin's calculation of the denudation
of the Weald and John Phillips' estimate from the erosion of the
Ganges Basin. Although not unique, Darwin's limited approach using
a single denuding action and a single geological formation was not
typical of the geologists' methods. Phillips' use of data from throughout the world and his emphasis upon subaerial rather than marine
denudation were more representative. Briefly, Phillips' method involved determining the rate of denudation for a given area of land for
which data was available, testing his result against any other available
data, and then treating it as typical of the world as a whole. He then
assumed essential uniformity of geological action in the past and present and an overall balance between the area of land being denuded and
the area receiving sediment. From these data and assumptions he then
calculated the time required to build up the accumulated stratigraphic
deposits of the earth. All of Phillips' assumptions were rationally de-
fensible, they were in fact the simplest available to the method, but they
were not the only ones which could be justified by the available evidence. Indeed, within the framework of Phillips' basic approach, the
variety of possible assumptions provided the geologists with remarkable flexibility which they exploited freely during the next three
decades.

T. Mellard Reade (1832-1909), for example, began with a method
similar to Phillips, but because of a different choice of assumptions and
parameters, arrived at a far different result. The most obvious difference between the two methods lay in the fact that whereas Phillips had
concentrated on the rate of mechanical denudation as measured by the
solids suspended in the Ganges River, Reade concentrated upon chemical (or solvent) denudation as estimated from the soluble minerals
contained in the rivers of England and Wales. In his initial calculation,
however, Reade found that he could not neglect mechanical sedimentation altogether, and thus he tried to make an appropriate allowance
for the deposition of both mechanical and chemical sediments. For this
purpose he drew heavily upon the extensive data made available by
studies of the sedimentary load carried by the Mississippi River. Reade
also introduced several other corrections which Phillips had ignored,
including corrections for volcanic sediments and marine denudation,
but the major quantitative difference between the two calculations
arose from the way in which he chose to estimate the volume of

sedimentary material in the earth's crust. Starting from the common assumption that the continents and sea bottoms have alternately risen and sunk, he concluded that there was no basic difference in the earth's crust under the exposed land masses or beneath the seas, and on the basis of admittedly sketchy evidence, he estimated the average depth in both places at ten miles. He then made the implicit assumptions that the exposed areas of the continents undergoing denudation have always remained nearly constant and the rates of denudation essentially uniform, and proceeded to calculate the time necessary for this denudation to produce a quantity of sediment equivalent to a crust ten miles thick covering the whole earth. The initial result of the calculation indicated that the time required would be about 526 million years, a figure that Reade regarded only as an approximation. He did insist, however, that it was an approximate *minimum*.[16]

Three years later Reade's concern with solvent denudation led him to attempt a further improvement in his calculation. Without changing his basic method, he concentrated his analysis upon a single sedimentary material, limestone. He hoped in that way to be able to ignore mechanical sediments entirely and to deal exclusively with chemically soluble carbonates and sulfates of lime. His aim was to refine his procedures by reducing the number of parameters considered. Instead he found it necessary to increase the number of assumptions and estimates needed for calculation. For example, he now had to estimate not only the thickness of the earth's crust, but the proportion of it made up of limestone as well. He also made other corrections such as increasing the estimated depth of the earth's crust from ten to fourteen miles. But each new set of assumptions seemed to balance out, and ultimately his new result of 600 million years differed little from his original estimate.[17]

Reade's calculations appeared between 1876 and 1879 well after Kelvin's conclusions had become widely accepted, and he clearly hoped to reverse the trend. A successful architect and engineer as well as a prolific geological amateur, he had been drawn like Kelvin to the physical side of geology. He was fascinated by the problems of the earth's interior temperature and rigidity, but unlike Kelvin he was not willing to disregard the evidence of geology. Thus, throughout the seventies he vigorously opposed Kelvin's arguments and denied the validity of his conclusions.[18] "In the present state of science," he declared in 1876, "it defies calculation to reach a maximum beyond which we can say the age of the earth does not extend."[19] But he stood against the tide and by 1893 he too capitulated.

Forced by the evidence of the *Challenger* expedition to abandon the assumption that the ocean bottoms were sedimentary, Reade fell back upon the more conventional assumption that the area being built up by sedimentation was equal to the area being denuded to provide the sediment. He also drastically reduced his estimate of the depth of the earth's sedimentary crust to bring it into line with the mass of new evidence. The major innovation in the new calculation, however, was the introduction of an assumed "effective area" of denudation (ap-proximately one-third the true area) by means of which he hoped to compensate for the cyclic reuse of sedimentary materials in successive strata. But though this assumption served to increase Reade's estimate of time by a factor of three, his result was still far below his earlier estimates. This time, he found only 95 million years separating the present from the beginning of Cambrian.[20] It is indicative of the sometimes indirect manner in which Kelvin's influence had spread that although Reade made no reference to any of the physical arguments, he happily noted the agreement between his 95 million years and Geikie's 100 million years. The passage of two decades had evidently obscured the fact that Geikie's figure had been drawn directly from Kelvin and not calculated from the evidence of geology.

A more interesting approach, both in the conception of the argu-ment and in the way assumptions and preconceptions were imposed on the evidence came from Samuel Haughton (1821-1897). Although for thirty years the professor of geology at Trinity College, Dublin, Haughton's eclectic—and one might add, eccentric—interests and his passion for numerical calculations kept him from sharing the biases of his more conventional colleagues. As early as 1862, for example, he was willing to entertain the possibility that Kelvin's 100 million years might be sufficient for geology, and he denounced his fellow geologists for indulging in the "wildest and most extraordinary statements and speculations." The geologists, he went on, "speak of the enormous lapse of time requisite for the formation of exceedingly small quantities of rock in a manner that would almost make us suppose that some miraculous agency was at work to retard the progress of the formation of those rocks."[21]

At the time, Haughton believed that geological knowledge was con-fined to relative ages and that any reference to absolute dates was mere speculation. When, two years later, he turned to his own attempt at calculating time, however, he completely reversed his view. Working from the assumption that past changes in climate have followed the gradual diminution of the earth's internal heat, he adopted a very different measure of the rate of secular cooling from the one developed by Kelvin, and found that 2,298,000,000 years had elapsed just between the time of the formation of the oceans and the beginning of the Tertiary period. Such enormous periods of time, he admitted, were "practically infinite" since they were "so great as to be inconceivable by beings of our limited intelligence."[22]

After arriving at this essentially uniformitarian conclusion, Haughton abandoned his "geological calculus" until 1878 when once again he found it necessary to reverse his views.[23] At first, however, he seemed to be merely refining the approach developed in 1864. He once again began with the assumption that climatic changes had followed the secular cooling of the earth, and by comparing fossil samples from the arctic with analogous forms presently existing in temperate and tropic climates, constructed a scale correlating past arctic temperatures with successive geological periods. He also drew upon physical measurements of the rate of cooling of solid bodies to determine the relative durations of the periods thus defined. The only new element, other than substituting arctic fossils for the English fossil mollusks used earlier, was the use of a sedimentation rate to strata thickness ratio as his measure of absolute time. But it was there that he found it necessary to introduce a method of procedure that dramatically illustrates the influence of Kelvin's limited time scale.

Haughton's periodization by temperature placed the beginning of geological time at the point when arctic temperatures first reached 212°F. Water could then start to condense and meteorological cycles necessary for geological change could begin. Life, however, could only begin after the surface had cooled below the coagulation temperature of albumen, or 122°F. Below that, Haughton's scale of fossil comparisons placed the beginning of the Triassic period at 68°F and the Miocene epoch at 48°F. The present, which he believed to be a period of prolonged thermal equilibrium, was assumed to correspond to an average arctic temperature of 0°F. Turning next to measurements of strata thickness and the rate of cooling of solid bodies, he concluded that the relative durations of his three pre-Miocene periods were approximately equal and that a similar period had been necessary for the temperature to drop from 48°F to 32°F. He did not believe his calculations could be accurately carried below 32°F because of the approach to thermal equilibrium, but he assumed that the subsequent interval must have been very long. The amazing result of this unique periodization was that:"A greater interval of time now separates us from the Miocene tertiary epoch than that which was occupied in producing all the secondary and tertiary strata, from the triassic to the miocene epoch."[24] In other words, he had arrived at either a vast expansion of post-Miocene time or an incredible contraction of the Paleozoic, Mesozoic, and early Cenozoic eras. And in either case it meant ignoring large bodies of conflicting evidence from stratigraphy.

Despite his singular periodization Haughton had thus far proceeded methodically, but when he began to convert his results into terms of years his mathematical juggling was little short of a swindle. He seemed at first to be following the conventional method of determining the ratio of strata thickness to the rate of denudation. With a slight modification, he adopted Geikie's widely accepted estimate that the average rate of continental denudation has been one foot in 3,000 years; but by making the assumption that the sediment thus produced is deposited over the whole ocean floor, he reduced the rate of sedimentary buildup by the ratio of land area to sea area, or by a factor of three. He then further prolonged the process of sedimentation by choosing an abnormally large estimate of strata thickness (33 miles as compared to Reade's 10 to 14 miles and much smaller estimates by Geikie and others). Without further modification, Haughton's data would have given a duration of 1.5 *billion* years for pre-Miocene time—an estimate closely in accord with the result he had obtained fourteen years earlier—but instead he arbitrarily and without explanation reduced his result by a factor of ten to 153 *million* years.

As strange as this procedure may seem, it appears even stranger when viewed against other parts of the calculation. In adjusting Geikie's estimate of denudation to account for the land/sea ratio, for example, Haughton casually converted an approximate average which Geikie had rounded to one foot in 3,000 years to an exact figure expressed as one foot in 8,616 years. The implied precision of four significant figures was far beyond anything Haughton's data could possibly justify, and yet he compounded the felony by keeping all four figures even after arbitrarily reducing this result by a factor of ten. The whole procedure is frankly inexplicable, since Haughton was certainly mathematician enough to recognize the inconsistency in what he was doing. It is extremely significant, however, that although Haughton's calculation was cited repeatedly during the next two decades and his estimate of 153 million years was regarded as one of the proofs that the earth's age is limited, these inconsistencies were totally ignored.

It seems obvious that one of the major reasons behind Haughton's manipulations was the desire to bring his conclusions into line with the growing body of opinion that a few hundred million years was all the time that could be allowed for terrestrial history. But another, perhaps equally significant part of his dilemma arose from the same problem that had caused Charles Darwin so much discomfort, that is, how to reconcile biological evolution with the physicists' short time scale. Haughton's intellectual sympathies were clearly on the side of evolution and vast geological ages, and his skepticism toward his own results is clearly apparent in his pointed disavowal of any belief that denudation actually proceeded ten times faster in the past than in the present. The weight of opinion was against him, however, and he felt compelled to force—the word is not too strong—his results into partial agreement with those of his fellows. He was faced in consequence with an inhospitably brief time span which he adjusted to provide the time which he believed necessary to account for the evolution of the complex array of post-Miocene mammalian species. His analysis of the earth's cooling made such an adjustment possible even though it was contrary to the evidence of stratigraphy; but he left little doubt that it was the biological problem that had played the critical role (italics his):

> The enormous interval of time that separates us from the miocene epoch affords ample opportunity for the development of gigantic mammals, which are commonly supposed to have suddenly made their appearance on all our continents and have disappeared as suddenly.[25]

Faced with the seemingly irreconcilable demands of physics, geology, and biology, Haughton sought compromise even at the expense of disregarding his own data, but the result was neither smooth nor satisfying.

A smoother but no less interesting attempt at compromise appeared three years later in Alfred Russel Wallace's *Island Life*. Haughton's biological problem centered around the necessity of accommodating the evolution of recent mammalian species within a limited time scale. Wallace, on the other hand, was concerned to provide sufficient time for the Precambrian evolution of primitive life. The two therefore faced very different problems and yet taken together their efforts provide an excellent illustration of the range of possible solutions obtainable from the same basic data, and of the importance of preconceptions and assumptions in interpreting such data.

Far more than Haughton, Wallace was convinced of the necessity of working within the chronological limits set by Kelvin. As he made clear at the outset:

> Sir William Thomson concludes that the crust of the earth cannot have been solidified much longer than 100 million years (the maximum possible being 400 million), and this conclusion is held by Dr. Croll and other men of eminence to be almost indisputable. It will therefore be well to consider on what data the calculations of geologists have been founded, and how far the views here set forth, as to frequent changes of climate throughout all geological time, may affect the rate of biological change.[26]

Wallace had attempted the reconciliation between Kelvin's views and evolution as early as 1869, but when he came to review it for a more complete presentation, he found that ten years and much more extensive data had considerably altered some of his premises. Among other things, he abandoned the idea, drawn from Croll's original glacial theory, that the continental ice sheets would alternate from hemisphere to hemisphere every ten thousand years. He leaned instead toward Lyell's belief in the simultaneous glaciation of both hemispheres. He still accepted the hypothesis that pronounced orbital eccentricity was necessary for the onset of an ice age, however, and the net result of his modified theory was that his supposed periods of rapid migration and accelerated species change were limited to the periods near the beginning and end of the glacial eras. Such periods, he was sure, had been quite numerous enough to account for rapid species change without recourse to any mechanism beyond natural selection.

Having therefore successfully preserved the basic point in his earlier argument, namely, that due to an extended period of low orbital eccentricity the present rate of species change is untypically slow, Wallace abandoned his original chronometer based upon Lyell's cycles of species change, and turned to the more conventional measurement of strata. Like most of his fellows, Wallace took his data from published sources, but what is particularly interesting, he chose the same basic parameters that Haughton had used—Geikie's one foot in 3,000 years as the rate of denudation and Haughton's 33 miles (or 177,200 feet) as the thickness of strata. He changed one critical assumption, however, and arrived at a radically different result.

The difference in the two calculations lay in how the two men approached the distribution of sedimentary material. Whereas Haughton had assumed that sediment was deposited uniformly over the whole sea bed, Wallace argued that the area of sedimentation must be confined to a narrow coastal strip averaging only about thirty miles in width. Thus instead of sedimentation occurring three times more slowly than denudation, Wallace's assumption would have it proceed nineteen times faster. Even without further corrections this change would reduce Haughton's results by a factor of 57, and Wallace need go no further. Having accounted for both the relative stability of modern species and the acceleration of biological change in the past, he had no need for Haughton's prolongation of recent epochs. Nor did he need to reduce his quantitative results arbitrarily by a factor of ten. Given his assumed conditions, the conventional ratio of strata depth to the rate of denudation would yield an estimate of only 28 million years since the beginning of the Cambrian.[27] And thus once again he had provided both for a long Precambrian period of evolution and for the limits Kelvin had imposed on geology. Compared with Haughton's billion and a half years from the same basic data, this result is truly astounding, but it is perhaps even more revealing when compared to Wallace's own earlier figure of 24 million years drawn from an entirely different method of calculation. Probably no other set of results better illustrates the importance of the geologists' assumptions and the consequent tenuousness of their results.

Each of the calculations discussed above was in some way unique, yet despite individual differences they all contributed to a consensus of geological opinion that grew steadily stronger throughout the seventies and eighties. By the early eighties, in fact, innovative work on the age of the earth had practically ceased. Few geologists were interested enough to attempt major revisions in hypotheses which seemed so consistent in the apparent agreement of their results. The fact that often this agreement was more apparent than real was lost as the numerical results of Kelvin, Croll, Haughton, and others were repeated monotonously, while the assumptions upon which their arguments were based faded into the background. Indeed, not only did this concentration upon numerical results obscure the tenuous nature and mutual contradictions of many of the calculations, it even concealed the fact that some of the estimates cited were not based upon calculation at all. Thus two of the more frequently mentioned "authorities" were Geikie and James Dwight Dana, although neither had done more than endorse Kelvin's results.[28] Occasionally, of course, an individual

hypothesis was challenged, but the general belief in the validity of the
established methods and in the approximate accuracy of the accepted
results remained unshaken.

This temporary hiatus ended abruptly in the nineties when, for a few
years, geologists poured out a concentrated torrent of new calculations
dealing with the age of the earth. Produced as a result of renewed
pressure from physics to reduce the geological demands for time still
further, the new calculations tended to be polemical rather than innovative. Few of them brought anything original to the discussion, and
most were mere rehashes of arguments presented twenty years before.
They were, of course, modified somewhat to take into account new
data, but they still provided predictable results. What was significant
about the renewed interest in the question of time was the fact that for
the first time it was international in scope. After nearly half a century
the British monopoly was broken as Americans and a few Europeans
entered the fray, but the effects of Britain's long dominion remained
clearly apparent.

The most significant British contribution to the renewed discussion
came from Archibald Geikie who, much as he had twenty-five years
earlier, set the pattern for discussion for the rest of the decade. Speaking as president of the British Association in 1892, Geikie praised
geology for the tremendous advance it had made during the preceding
century and paid homage to Kelvin for rescuing it from the errors and
extravagance of relying upon unlimited time. He then went on to
attack Kelvin and Tait for the later restrictions that they had attempted
to impose, and attempted to refute them on the basis of geological
evidence. And at long last, after years of being cited as an authority for
the geological estimate of 100 million years, he made his own calculation of time. The calculation itself was hardly significant. It followed
Phillips' original method but with modified data that produced a
Kelvinesque range of 73 million to 680 million years.[29] The important
thing was that Geikie, who as much as any one else had been responsible for establishing the conviction that the earth's age was limited, had
used the prestigious pulpit of the presidency of the British Association
to defend the geological *status quo*.[30]

Among continental geologists, interest in the earth's age remained
minimal until the end of the century. They still looked upon the
problem as primarily a British preserve. Thus as late as 1899, Karl von
Zittel's *Geschichte der Geologie und Paläontologie* devoted little more than a
page to the question of time and mentioned only the physical results of

Kelvin, Helmholtz, and Clarence King.[31] The British, on the other hand, quickly seized upon the results from the only quantitative calculation of the earth's age to appear on the continent in nearly forty years. Their warm reception is understandable since except for language, Albert de Lapparent's paper of 1890 might have been written by one of themselves. De Lapparent's method of approach was drawn indirectly from Phillips; his operating assumptions were very nearly those of Wallace; and his data was drawn primarily from such British sources as Geikie, Croll, and the *Challenger* expedition. It is hardly surprising, therefore, that he found the age of the stratified rocks to be between 67 million and 90 million years.[32] One such calculation admittedly cannot be used as a measure of the whole of European opinion, but it is certainly an indication of the range of British influence. It is also significant that de Lapparent pointed with obvious satisfaction to the fact that his results fell within the 100 million year limit allowed by Kelvin.

The most concentrated outpouring of new chronological calculations, however, came from the United States. After years of exploration, the pioneering surveys of the American West had produced a wealth of data from which to reconstruct the earth's past history. Yet when American geologists came to attack the problem of absolute time, they looked inevitably toward British models, and their results mirrored British results. In successive issues of the same journal, for example, Clarence King (1842-1901) attempted to refine Kelvin's arguments and lower the acceptable limits of the earth's age, while Warren Upham (1850-1937) reviewed the various geological estimates and came down firmly in support of Geikie.[33] A few months later Charles D. Walcott (1850-1927), like King and Upham a pioneer in the U.S. Geological Survey, produced a calculation which in method, spirit, and results was indistinguishable from any of a half dozen of its British counterparts. His conclusion that "geologic time is of great but not of indefinite duration . . . that it can be measured in tens of millions, but not by single millions or hundreds of millions of years," expressed the prevailing opinion on both sides of the Atlantic.[34] For a year or so around 1893, American geologists virtually dominated the discussion of geological time. They reviewed previous arguments, made independent calculations, and condemned the restrictions of the physicists; but in the end, their task was one of summation rather than innovation.

Emphasis upon the apparent agreement among the quantitative

results obtained from the various calculations became a dominant theme in the discussion of time during the nineties. The arguments themselves were condensed in these summaries until only the numbers remained while the assumptions upon which they were based and the inconsistencies which they concealed were lost from sight. Little was left of the uniformitarianism of the mid-century except the belief in the present as the key to the past. The same process of repetition and condensation that had produced agreement on the quantitative limits of time had worked to create a myth that the great generalizations of the mid-century were vague, speculative, and somehow unscientific. It became almost *de rigueur* to pay homage to Kelvin for freeing geology from the excesses of uniformitarianism even while attacking the later restrictions that he was still attempting to impose. Above all, geologists no longer felt obliged to submit meekly to the demands of physics. They had developed their own mensural and quantitative techniques which provided consistent and convincing results. There were, of course, exceptions to the prevailing consensus, but certainly Warren Upham spoke for the majority of his colleagues on both sides of the Atlantic when he wrote:

> Reviewing the several results of our several geologic estimates and ratios supplied by Lyell, Dana, Wallace, and Davis, we are much impressed and convinced of their approximate truth by their somewhat good agreement among themselves, which seems as close as the nature of the problem would lead us to expect, and by their all coming within the limit of 100,000,000 years which Sir William Thomson estimated upon physical grounds. This limit of probable geologic duration seems therefore fully worthy to take the place of the once almost unlimited assumptions of geologists and writers on the evolution of life, that the time of their disposal has been practically infinite. No other more important conclusion in the natural sciences, directly and indirectly modifying our conceptions in a thousand ways, has been reached during this century.[35]

Physics and the Limitation of Time

While the geologists were adjusting to Kelvin's original limits on the earth's age, a handful of physicists and astronomers seemed intent upon reducing those limits still further. Starting usually from one or another of Kelvin's initial arguments, the "physicists,"[36] including Kelvin himself, came to the conclusion that he had been far too generous in his original allowance for time, and proposed limits of 50 million or 20 million or even 10 million years. This view first attracted attention when it appeared in Tait's anonymous article in the *North British Review* in 1869. But its real prominence began with a series of public lectures delivered by Tait in 1874 and published two years later as *Lectures on Some Recent Advances in Physical Science.*[37]

It was entirely characteristic of Tait that in championing Kelvin's conclusions he pushed them to limits not yet envisioned by their creator. In 1874 he had been working closely with Kelvin for nearly fifteen years and thus was in an ideal position to understand the importance that the older man attached to his calculations. But Tait's unswerving loyalty to his friends and his absolute faith in the validity of mathematical physics had more than once led him to take dogmatic and sometimes extreme positions in public debate, and so it was in the present instance. Delivered extemporaneously from a few notes, his lectures contributed nothing new in the way of evidence or calculations. Instead, they reflected his opinions rather than the rigorous development of an argument. Towards geology his tone was polemical and sometimes frankly contemptuous. He denounced uniformitarianism as "totally inconsistent with modern physical knowledge,"[38] and cavalierly dismissed Lyell and Darwin with the remark:

> So much the worse for geology as at present understood by its chief authorities, for, as you will presently see, physical considerations from various independent points of view render it utterly impossible that more than ten or fifteen millions of years can be granted.[39]

Such high-handed treatment naturally generated antagonism, but the frankness and spontaneity of the lectures made them extremely popular, and they exerted a considerable influence.

Tait's comments on the age of the earth were based entirely upon Kelvin's arguments. He examined them each in turn and admitted that the results drawn from them depended upon the twin assumptions that science knew all of the physical laws now in operation and that these laws have remained unchanged since the earth was formed. He saw no reason to doubt either assumption. The only possible objection that he saw to Kelvin's calculations was that they were too generous in the breadth of possible error that they would allow. Further calculations of the earth's secular cooling, he argued, would reduce the time since its solidification to 10 million or 15 million years at most, while the evidence from tidal friction pointed to an even shorter time, probably *less* than 10 million years. Contrary to most of his contemporaries, including Kelvin himself, Tait viewed the sun's heat as the least restrictive of Kelvin's arguments, and at one point conceded that the nebular theory could account for as much as 100 million years of solar radiation at the present rate. What he gave with one hand, he took with the other, however, and in the next lecture he again asserted the belief—unsupported by any calculation—that the sun had actually heated the earth for no more than 15 million or 20 million years.

Even Kelvin, who had already shown his willingness to reduce his original estimates, found Tait's figure excessively small. Other commentators were still less enthusiastic. Darwin regarded Tait's conclusions as "monstrous," and wrote of his hope that they would soon be answered.[40] One such answer came in 1878 from Maxwell H. Close, the president of the Geological Society of Ireland. Adopting an approach that was to become typical two decades later, Close avoided Tait's emphasis upon the certainty of the principles and methods involved in Kelvin's arguments and stressed instead the uncertainties in his data and assumptions. Confining himself strictly to physical considerations, Close pointed out that the physicists themselves disagreed over such fundamental questions as the nature of nebulae and the constancy of gravitation in interstellar space, and he charged that Kelvin's own studies of the tides had weakened the arguments for terrestrial rigidity and for dating the earth's age from tidal retardation. Furthermore he argued that the bulk of the data employed by Tait and Kelvin came not from measurement but from uncertain estimates or unproven assumptions, and therefore could not possibly support the degree of precision claimed by Tait's results.

Few geologists had the effrontery to take on the physicists on their own ground as Close had done. By the seventies most of them were willing to accept the fundamentals of Kelvin's position even though appalled by Tait's proposed restrictions. C. Lloyd Morgan's response to Tait may justly be taken as representative of the position of the rank and file of geologists at the time. Like Close, Morgan challenged the precision that Tait would claim from uncertain data, but beyond that he made no attempt to dispute the physicists' arguments. He accepted as proven the belief that physical considerations limited the earth's age

to between 10 million and 100 million years. But if Kelvin and Croll
had saved geology from the excesses of unlimited time, Tait was guilty
of excess in the other direction. In Morgan's opinion, 100 million years
or even half that amount would be sufficient for the demands of
geology, but he was equally sure that the evidence of biology decisively
refuted any lower limit.[42] In short, compared with Tait, Kelvin's 100
million years seemed generous indeed; and for Morgan and most of his
fellows the lasting effect of Tait's pronouncement was to solidify opinion around Kelvin's original result as reasonable and moderate.

Tait was not entirely without allies, however. The American astronomer Simon Newcomb (1835-1909) independently reviewed
Kelvin's arguments during the late seventies and arrived at a numerical
result almost identical with Tait's. As one might expect of an astronomer, Newcomb emphasized the problem of the sun's heat and
dismissed the other calculations as either less important or less exact.
The first part of his argument followed Kelvin's very closely as he
eliminated all the possible sources of solar energy except gravitation.
In his actual calculations, however, he abandoned Kelvin's reliance
upon the initial heat of collision to concentrate upon the energy released by the gradual contraction of the sun's mass. Based on the
present volume of the sun, Newcomb determined that gravitational
contraction could account for 18 million times its present annual heat
loss. Like most of his contemporaries, he ignored the sticky question of
how much of this energy has been expended in the past and how much
remains for the future, and turned instead to the assumed secular
cooling of the sun to limit his calculations still further. If one considers
only the time during which solar radiation could support life on earth,
that is, the time during which it would support a terrestrial surface
temperature between the boiling and freezing points of water, that
period must have been much shorter than the sun's total age. Again
ignoring the problem of what part of this period lies in the future and
making no further calculations, he estimated that these conditions
have probably not existed for longer than 10 million years.[43]

Looking only at their numerical results, Newcomb and Tait seem to
be in close agreement, and yet this apparent agreement illustrates the
profound ambiguity concealed in the way the age of the earth had
come to be discussed. Rather than reinforcing each other as most
contemporary commentators assumed, the arguments of Newcomb
and Tait taken together expose the uncertain nature of their assumptions. Tait, a physicist, saw the age of the sun as the least restrictive of all

of Kelvin's arguments. The nebular theory, he believed, was "thoroughly competent to explain 100 million years of solar radiation at the present rate, perhaps more."[44] That he believed the probable age of the earth and sun to be less than 100 million years was due to what he regarded as the conclusive arguments from the earth's secular cooling and the retardation of the tides. Newcomb, on the other hand, regarded the sun's heat as the conclusive argument. The mass and energy of the sun permitted the calculation of exact results whereas the depth and rate of cooling of the earth's crust were both unknown and beyond meaningful calculation. Solidification of the earth's crust, he argued, might have occurred 100 million or even 1,000 million years ago. Indeed, the whole principle of the earth's secular cooling might be vitiated if terrestrial contraction were still generating more heat than is lost by radiation.[45] Thus the superficial agreement between Newcomb and Tait collapses. Although their numerical results were remarkably similar, their fundamental positions were mutually contradictory. And yet it is particularly significant that this contradiction was either ignored or overlooked by Tait, Newcomb, and their contemporaries; and for the next two decades the correspondence between the numerical results was repeatedly cited as *proof* that the methods and the order of magnitude of the results were both correct.

Newcomb and Tait had done little more than elaborate on Kelvin's basic arguments, but occasionally entirely separate problems arose which seemed to shed light on the earth's age. Of these, one of the most interesting and influential was George H. Darwin's (1845-1912) theory of the origin of the moon. Even here, however, Kelvin's influence remained significant. As the second son of the famous naturalist, Darwin was, of course, acutely aware of the importance of Kelvin's chronology. Moreover, early in his own career he had formed a close intellectual and personal bond with the older physicist as a result of Kelvin's enthusiasm for his first major paper—a physical analysis of the effects of geological change upon the earth's axial rotation. Thus it was as a kind of informal protégé that Darwin responded to Kelvin's encouragement and began the investigations that in 1879, led to his theory of lunar genesis.

Orginally Darwin had no intention of becoming involved in the problem of the moon's origin nor that of the earth's age. He set out to make a mathematical investigation of the properties of a theoretical viscous spheroid, using the example of the moon's influence upon the rotation of the still molten earth merely as a case study in theoretical dynamics.[46] Quite naturally he began with the same postulates used by Kelvin. He assumed that the moon acted through the friction of the tides—in this case, earth tides in the primordial molten mass rather than oceanic tides—to slow the earth's velocity of rotation, and that the earth reacted to increase the moon's moment of motion, its orbital radius, and its period of revolution. The velocities of the earth's rotation and the moon's revolution must therefore have slowed together but at different rates. None of this was at all novel until Darwin asked the critical question: under what conditions would the length of the month and day be the same? Making the assumptions necessary for calculation, he found that identical periods would occur at 55.5 days and at 5 hours and 36 minutes. He dismissed the former case as a future state toward which the earth-moon system is progressing, but the latter case seemed to him to present significant possibilities for geology and cosmology.

One of the consequences of the inverse square law of gravitation is that the period of rotation of a satellite is proportional to the three-halves power of its distance from the mother planet. In the case of the earth-moon system, Darwin found that a month of 5 hours and 36 minutes would correspond to a lunar distance of only 10,000 miles. As even more suggestive, he found that reducing the month by only twenty minutes would reduce the lunar distance to 6,000 miles. "This small distance," he observed, "seems to me to point to a breaking up of the earth-moon mass into two bodies at a time when they were rotating in about 5 hours; for of course, the precise figures given above cannot claim any great exactitude."[47] There were major problems involved in the hypothesis, most notably that of accounting for the moon's expulsion from the earth at a time when the centrifugal force of the system would be less than half that necessary to produce the fissure. If, however, the actual cause of dissociation were neglected pending further evidence, Darwin's calculations indicated that the tides acting on the still viscous earth could produce months and days of the present duration in "something over 54 million years."[48]

A second part of Darwin's analysis seemed to reinforce the result from the first. Searching for the possible effect of moon-generated earth tides in changing the obliquity of the ecliptic, he found that if he again assumed the initial condition of a day and month of the same length, tidal drag could account for the present tilt of the earth's axis in something less than 57 million years.[49] Even allowing for the uncertainties inherent in a necessarily complex and indirect calculation, such a correlation of results seemed too incredible to be·mere coincidence.

Two points in Darwin's calculations require careful attention, however, particularly since they were usually overlooked when his results were compared with those of Kelvin. First, his results gave only *minimum* estimates of the time necessary to establish present lunar and terrestrial conditions. They in no way limited the possibility of much greater maximum results. Darwin was well aware of this distinction, but unfortunately failed to emphasize it strongly enough. In fact, at one point he specifically noted that his results fell well within Kelvin's limits for the earth's age. Secondly, Darwin's calculations were based upon earth tides acting in a viscous, non-rigid globe, while Kelvin's depended upon oceanic tides acting over the surface of a rigid earth. The distinction is critical. Indeed, in an often overlooked part of his argument Darwin took up the problem of oceanic tides only to dismiss them from further consideration because he found that they would increase the length of the month by only one day in 700 million years and would not perceptibly alter the length of the day at all.[50].

Such considerations had little bearing on the theory of lunar genesis since, as Darwin pointed out, his calculations would remain valid for the originally molten earth even if it were now completely rigid. They did, however, have a direct bearing upon the applicability of his results to the age of the earth. If the retardation of the motion of the earth and moon depend upon earth tides, if oceanic tides have virtually no effect, and if the earth is not truly viscous, all of which Darwin accepted, then most of the 54 to 57 million years necessary to establish the present earth-moon system must have taken place either before the earth solidified or while a fluid interior remained under a solid crust. Yet if Kelvin were correct about the earth being perfectly rigid, all of the effects investigated by Darwin must have been prior to the formation of a rigid crust. Darwin recognized the possibility of this objection, but as he was unwilling to concede that the whole process was pregeological, he found it necessary to reject Kelvin's dogma of a perfectly rigid earth. He suggested instead that the earth must possess some degree of plasticity, enough so that it would be slightly viscous with respect to tidal pull and yet appear sufficiently rigid with regard to sheer stress to be called "solid" in ordinary language. Under such conditions, Kelvin's argument from tidal retardation would collapse, but as Darwin pointed out, it would be more than compensated for by the acceleration of geological processes caused by larger and more frequent tides, more rapid alternation of day and night, and more violent winds and storms.

Within a year (1880), Darwin had reversed his position on oceanic tides. Further calculations had convinced him that while it was "not impossible" that the action of earth tides had been responsible for the early changes in the earth-moon system, the more recent changes had been due to the friction of ocean tides. This reversal had no effect on the basic theory of lunar formation, since all of the effects attributed to earth tides would also follow from oceanic tides if *much more time* were allowed. Of course it was now necessary to postulate a sufficient time for the new hypothesis, but this presented no real difficulty. Darwin's original 54 million and 57 million year figures had been minima only, and despite his ties to Kelvin, if more time were needed—even several hundred million years—he was ready to accept it.[51]

Darwin's theory of lunar genesis struck a responsive chord in the minds of scientists and laymen alike. Although the mathematical complexity of his reasoning was beyond the grasp of most laymen and most, if not all, geologists, a shorter, less technical version of his conclusions made them comprehensible to any interested reader.[52] Unfortunately, in the popularized version many of the assumptions involved in his mathematics were necessarily lost. Also since the popular version was based upon Darwin's original paper, it did not contain the revisions of the second article. Thus too little emphasis was given to the fact that the figures 54 million and 57 million years for the age of the moon represented only possible minima and in no way limited the maximum age of the earth-moon system. Instead these figures passed into the literature as still another proof that the earth *must* be less than 100 million years old.

After Darwin's lunar theory, physics and astronomy contributed nothing new to the problem of geological time for more than a decade. Kelvin continued to whittle away the upper limits of his estimates, thus giving rise to an occasional small debate, but nothing really innovative appeared until 1893. When it did appear, the new physical approach to time was the work of an American geologist, Clarence King (1842-1901), rather than that of a physicist. King, who had served briefly as the first director of the United States Geological Survey, was a sometimes brilliant but erratic thinker. At home both in the advanced intellectual circles of the east and in the geological field camps of the west, his ideas frequently differed from those of his more conventional colleagues. Like Kelvin, he was a catastrophist who rejected the idea that the relative uniformity of the most recent geological periods could be used as a standard for all past geological time.[53] And as director of the Geological Survey, he had been among the first to advocate the extensive use of physical and chemical laboratory procedures in the investigation of geophysical problems. It is hardly surprising, therefore, that his approach to the question of the earth's age bore a greater kinship to the methods of Kelvin than to those of his fellow geologists.

By his own admission, King's approach owed much to Kelvin, but in a sense it began where the older man left off. Whereas Kelvin had merely assumed that the earth's secular cooling would account for its solid condition, King considered the problem too important to be left to assumption alone. He therefore elevated the question of the earth's solidity and rigidity to a position of paramount importance; and without questioning the overall validity of Kelvin's method, sought to determine whether or not his assumed conditions could actually have given rise to a completely solid earth. If not, King intended to find out just what the necessary conditions must be, even at the expense of sacrificing Kelvin's estimate of time.[54]

Starting from the premise that the fluid or solid state of the earth's interior must be determined by the balance between the opposing effects of heat and pressure, King saw his problem as one of establishing what the probable values of these two parameters must now be and what they must have been in the past in order for the earth to have remained completely solid throughout all geological time. Two factors seemed to simplify the problem. Laplace had already calculated the distribution of pressure with depth inside the earth, and Kelvin had argued convincingly that the earth's interior temperature must be uniform below the shallow depth affected by secular cooling. Taken together, these two hypotheses implied that the effects of heat could not possibly dominate those of pressure except in a very thin layer of the outer crust. Even in confining his analysis to this thin outer layer, however, King's problem remained formidable. His only fixed parameter was the present surface thermal gradient of the earth, while his two immensurable variables, the earth's initial temperature and its rate of cooling, were mutually dependent. He was consequently faced with a problem for which he could find no exact general solution and had to proceed laboriously by trial and error.

King lacked the mathematical sophistication needed to deal with his problem analytically. Instead he approached it graphically, constructing hypothetical profiles of the earth's internal condition corresponding to assumed values of his variable parameters. He then compared these to a similar profile of the rock *diabase* which, at his instigation, had been prepared experimentally by Carl Barus, and which he now took as representative of the earth's crust in general.[55] In this manner he determined that Kelvin's assumed conditions of 3900°C for the earth's initial temperature, and 100 million years for its time of cooling would produce a fluid layer 200 miles deep only 27 miles below the surface.[56] He further calculated that it would require 600 million years for an earth of Kelvin's assumed initial temperature to solidify completely. King found both alternatives unacceptable, and in a similar way eliminated several other pairs of possible initial conditions. It was easy enough to satisfy the condition of total solidity if sufficiently low initial temperatures and sufficiently short cooling times were assumed, but King's aim was to find maximum values. Ultimately he decided upon optimum values of 1950°C for initial temperature and 22 million to 24 million years for the time of cooling. These were a far cry from Kelvin's original conditions, but as King pointed out, the final result was in good agreement with his 20 million years for the age of the sun.

King had, in his own words, set out to extend Kelvin's conclusions "to a further order of importance."[57] What he accomplished was something altogether different. Kelvin himself welcomed King's result and saw in it one more justification for the downward revision of his time scale. The geologists, however, saw it in a very different light. By 1893 they had developed their own methods, quite apart from physics, for determining the earth's age; and in the process, had brought their demands for time into line with what they regarded as the legitimate limits set by physics. Now faced with new demands and an ever contracting time scale, they were no longer willing to capitulate meekly, even though the physics was wielded by one of their own number. The result was a period of renewed controversy which once again pushed Kelvin's views into the center of the debate.

REFERENCES

[1] For a valuable discussion of Lyell's views at the time of the first publication of the *Principles*, see: Rudwick (1970), *Strategy of Lyell's Principles*.

[2] Although Kelvin's results had been used by the anti-Darwinian press as early as 1862 (see: Ellegård (1958), *Darwin and the General Reader*, pp. 236-237), the full impact of his arguments was not felt until after the debate with Huxley.

[3] M.A. (1896), *Huxley's Address*. It is perhaps worth noting that in denouncing Evolutionism the author lists Kelvin as "one of the leading Evolutionists of the day." *Ibid.*, p. 277.

[4]Anonymous (1869d), *Geological Dynamics*.

[5]Anonymous (1869c) *Mathematics vs. Geology*.

[6]*Ibid.*, p. 12.

[7]Tait (1869), *Geological Time*. (For author, see: Haughton, W. (1966), *Wellesley Index*, I:963.)

[8]*Ibid.*, p. 217.

[9]Dawkins (1870b), *Geological Theory*.

[10]*Ibid.*, p. 42.

[11]The degree of agreement—often more apparent than real—between Kelvin's results and those drawn from other lines of evidence is startling. In 1924 G.P. Merrill surveyed some of the best known calculations of the earth's age including twenty estimates made by seventeen investigators between 1870 and 1900. Of the twenty, a few were quite low, but only four exceeded 100 million years and none exceeded 250 million years. Coincidence alone cannot account for this degree of agreement among calculations drawn from such diverse data. (See: Merrill (1924), *American Geology*, pp. 655-660.)

[12]Reade (1878a), *Geological Time*, p. 211.

[13]Sollas (1900), *Evolutional Geology*, p. 314.

[14]Ramsay, A.C. (1873-74), *Comparative Value of Geological Ages;* Prestwich (1895), *Collected Papers*, pp. 47-48; Dana (1880), *Manual*, p. 590.

[15]Geikie, A. (1871), *Modern Denudation*, p. 189; Sollas (1877), *Evolution in Geology*, pp. 1-3; Fisher (1874a), *Formation of Mountains*, pp. 62-63.

[16]Reade (1878a), *Geological Time*. Reade was careful to point out that his estimate of ten miles was not the average thickness of the earth's crust but rather a composite maximum. In this way he hoped to allow for successive cycles of erosion and sedimentation.

[17]Reade (1879a), *Limestone as an Index of Geological Time*.

[18]Reade (1878b), *Age of the World*.

[19]Reade (1878a), *Geological Time*, p. 232.

[20]Reade (1893), *Measurement of Geological Time*.

[21]Haughton, S. (1865), *Manual of Geology*, pp. 79, 83. From a lecture delivered in 1862.

[22]*Ibid.*, pp. 99-101. Quotation, p. 99. From an article in *The Reader* of 1864.

[23]Haughton, S. (1878), *Physical Geology*.

[24]*Ibid.*, p. 268.

[25]*Loc. Cit.*

[26]Wallace (1880a), *Island Life*, p. 205.

[27]*Ibid.*, pp. 205-228.

[28]Geikie's initial use of Kelvin's results is discussed in Chapter 3. Dana's discussion of time, such as it was, appeared in his influential *Manual of Geology*. (See: Dana (1880),

Manual, p. 391.) What Dana did was merely to suggest a way of accommodating the physical time scale to his own determination of the relative durations of the great paleontological eras. Time, he said, was *long,* but 48 million years *might* be long enough.

[29]Geikie, A. (1892), *Presidential Address,* pp. 187-188.

[30]Geikie's monumental article, "Geology," in the 9th edition of the *Encyclopaedia Britannica,* which subsequently became his often reprinted *Textbook of Geology,* devoted virtually its whole discussion of geological time to Kelvin's arguments and results. And though I have little direct evidence to support my contention, I suspect that Geikie's uncritical endorsement did much to reinforce the impression that Kelvin's results represented the dominant opinion in geology.

[31]Zittel (1901), *History of Geology,* pp. 168-169.

[32]Lapparent (1889-90), *De la mesure du temps.*

[33]King, C. (1893), *Age of the Earth;* Upham (1893), *Geologic Time.*

[34]Walcott (1893), *Geologic Time,* p. 368.

[35]Upham (1893), *Geologic Time,* p. 218.

[36]The term "physicists" is perhaps not entirely appropriate since the group included astronomers, physicists, and geologists with a physical bias. Because of the prominence of Kelvin and Tait, however, it was the term most frequently used during the late nineteenth century, particularly by the embattled geologists.

[37]Tait (1876), *Lectures on Physical Science,* especially pp. 150-175. The lectures were printed with little revision from a shorthand record of speeches given from a few notes. Thus they preserved the freshness and spontaneity as well as the arrogance and pugnacity of Tait's remarks.

[38]*Ibid.,* p. 164.

[39]*Ibid.,* pp. 167-168.

[40]Darwin, F. and Seward (1903), *More Letters of Darwin,* II:211-212. Letter to T. Mellard Reade, 9 Feb. 1877.

[41]Close (1878), *Geological Time.* Read before the British Association Meeting at Dublin, Aug. 1878.

[42]Morgan (1878), *Geological Time.*

[43]Newcomb (1878), *Popular Astronomy,* pp. 505-511.

[44]Tait (1876), *Lectures on Physical Science,* p. 154.

[45]Newcomb (1878), *Popular Astronomy,* pp. 511-513.

[46]Darwin, G. (1879a), *Precession of a Viscous Spheroid.* The discussion of the earth-moon system occupies only one-third of Darwin's long theoretical paper, but its importance is emphasized in the second half of his title ". . . and on the Remote History of the Earth."

[47]*Ibid.,* p. 104.

[48]*Ibid.,* p. 105.

[49]*Ibid.,* pp. 124-130.

[50]*Ibid.*, pp. 36-37, 124-126, 133.

[51]Darwin, G. (1880), *Secular Changes*, pp. 336-370.

[52]Darwin, G. (1879-80), *A Tidal Theory*.

[53]King, C. (1877), *Catastrophism and Evolution*.

[54]King, C. (1893), *Age of the Earth*.

[55]While working under King's direction Barus had carefully measured the properties of diabase under laboratory conditions of extreme temperature and pressure. King therefore believed that he had the best experimental evidence available for his calculations.

[56]Many geologists opposed Kelvin's demand that the earth be completely rigid and solid. They required some kind of molten or viscous layer beneath the crust to account for earthquakes, volcanoes, mountain formation, and so forth. King was thus implicitly rejecting the views of his fellow geologists and allying himself with Kelvin. It is worth noting that, only a few years earlier, King's fellow countryman, Joseph LeConte, had sought to show that a "fluid couche" beneath the earth's crust was possible even while the earth remained rigid enough to satisfy Kelvin and G. Darwin. Indeed, LeConte's method of approach, utilizing an analysis of the balance of pressure and temperature within the earth to show that the fluid layer could exist, was essentially the same as the one used by King to eliminate such a layer. (See LeConte (1889), *Interior Condition of the Earth.*) King was certainly aware of LeConte's work although he avoided any reference to it.

[57]King, C. (1893), *Age of the Earth*, p. 1.

V
Opposition and
Controversy

The opening session of the British Association meeting in 1894 promised to be a memorable one. The presidential address was to be delivered by a much respected former prime minister, the Marquis of Salisbury, and the responses were to be made by the newly elevated Lord Kelvin, England's most eminent living physicist, and by Thomas Henry Huxley, Darwin's aging bulldog. The meeting was to be held at Oxford for the first time since the legendary occasion when Huxley so dramatically rebuffed Bishop Wilberforce, and it was hoped that some of the old fire remained. No one knew quite what to expect from Salisbury. He was not a professional scientist, but his knowledge and sincere concern for science were widely recognized. When it came, the address was decidedly different from those usually delivered on such occasions. Instead of recounting the past accomplishments of science, he took as his theme science's continued ignorance of fundamental causes in nature. Ranging widely over the unanswered questions of science, he came at last to Darwin and evolution. There he freely acknowledged the importance of the doctrine of evolution and heaped praise upon Darwin himself, but rejected natural selection as an unproven and unprovable hypothesis. His brief against natural selection dwelt upon its conflict with the evidence of design in nature and upon the impossibility of its being demonstrated experimentally, but his only positive argument centered upon Kelvin's limitation of time. Although disclaiming participation in the controversy over the age of the earth, Salisbury asserted his belief in the validity of Kelvin's arguments, and he left no doubt of his conviction that Kelvin had *proved* that time was too short to accommodate all of the effects attributed to natural selection.[1]

When Salisbury sat down, two fabled opponents on the question of time were left upon the stage. The drama was heightened, moreover, by the fact that only months earlier both Kelvin and Huxley had republished the texts of their 1868-69 exchange.[2] Kelvin rose and delivered a gracious but conventional word of thanks to the speaker, and the stage was Huxley's. Those who knew him well had measured his growing impatience by the tapping of his foot, and later one of them described the scene: "When Huxley arose, he reminded you of a venerable gladiator returning to the arena after years of absence. He raised his figure and his voice to its full height, and, with one foot turned over the edge of the step, veiled an unmistakable and vigorous protest in the most gracious and dignified speech of thanks."[3]

The melodrama was over. For Huxley it was the last public appearance, and in his words, "the game was worth the candle."[4] But for Kelvin, it marked a turning point in the opposition to his theories of geological time. Huxley had resisted the temptation to tear Salisbury's address apart, but others had gone away disturbed that once again Kelvin's physics had been turned against the accumulated evidence of biology and geology. Thus the real impact of Salisbury's speech was not in what he said, but in the new and vigorous challenge that his words ignited. The discontent that had been building for two decades as the physicists progressively lowered their allowable limit of the earth's age was brought to a focus. And thanks to Salisbury, the geologists found an important ally from the side of physics itself.

Growing Discontent

There can be little doubt that Kelvin's results dominated the question of geological time during the quarter century preceding Salisbury's address, but it was never an unchallenged domination. No one seriously questioned the fact that the earth's age was limited, and most were willing to accept Kelvin's 100 million years, or at most his 400 million years, as adequate for the demands of geology. But despite this general concession, several important problems remained unsolved and an occasional doubter still spoke out. For one thing, a serious discrepancy continued to exist between Kelvin's own best estimates of the ages of the earth and sun. For another, his insistence upon a perfectly solid and rigid earth, upon which two of his calculations of time depended, was directly contrary to the many geological hypotheses that required at least some degree of terrestrial plasticity. Other objections, such as the failure of the physicists to agree among themselves about which of Kelvin's calculations was the more reliable, could be mentioned, but none of them were sufficient in themselves to challenge Kelvin's theories. What they could do, however, was to provide some justification for doubting the exactness of the physical estimates. And as Kelvin's estimates of time grew more and more restrictive and his pronouncements more and more dogmatic, this kind of doubt became more and more important.

Of all of Kelvin's arguments the most speculative and most restrictive was the one relating to the sun's heat. From the very beginning it aroused controversy and a certain amount of opposition. Even those who were not concerned about the earth's age were disturbed by the wantonness of nature's dissipation of the sun's enormous energy. No one seriously doubted that the only probable explanation for solar heat was a mechanical explanation, but several hypotheses were proposed whereby the cyclic reuse of solar energy might extend its usefulness. These hypotheses were sometimes ingenious, and for a time some of them exerted a limited amount of influence. Almost all of them, however, at some point or other contradicted the second law of thermodynamics. Not uncommonly these contradictions were overlooked in the sixties, seventies, and even in the eighties, but as the science of thermodynamics matured it could not continue to be ignored. Thus, although numerous alternatives were presented, at the century's end, the hypotheses of Kelvin and Helmholtz were still regarded as the best and the most probable available.[5] The repeated questioning had served a purpose, however. It had shown that Kelvin's assumptions were not uniquely possible, and it had given support to those who doubted the physicists' restrictions on the age of the earth.

The challenges had not been slow in coming. As early as 1863, James Challis (1803-1882), the Plumian Professor of Astronomy at Cambridge, took issue with the collision theory of solar heat. Turning instead to the aetherial theories then in vogue, he suggestsd that the sun and stars might continually exchange energy through the "dynamical action of the direct vibration of the aether."[6] Going still further, he speculated that not only might solar energy be conserved, it might actually be increased as the stellar undulations had "their effects multiplied to an incalculable amount by the reaction of the immense number of atoms" encountered in the bodies of the sun and stars.[7] The exact nature of the "reactions" envisaged by Challis was as vague as his understanding of thermodynamics, however, and despite the eminence of his Cambridge position, his speculations were generally ignored.

The problem of solar heat was too important to be ignored completely, however, and a few years later (1870) a much more elaborate hypothesis appeared which once again undertook to show how the sun could gather "from surrounding suns as much force as it radiates toward them."[8] *The Fuel of the Sun* was the work of an obscure chemist, William Mattieu Williams (1820-1892), and it reflected a desire to impose chemistry upon the dominant physical hypotheses of Helmholtz and Kelvin. Basically Williams proposed that the ultimate source of the sun's energy could be gravitational while the immediate source could be chemical. Assuming a universal atmosphere of varying density pervading all space, he postulated that the compression of this atmosphere in the vicinity of the sun would generate intense heat which would dissociate the sun's constituent elements. These would then burst from the sun's surface as solar prominences, expand for a time, and then be drawn back by gravitational attraction. Finally, upon returning to the solar surface, the stored energy of dissociation would be released gradually through the recombination of the chief constituent elements, oxygen and hydrogen. By this curious circular reasoning, Williams had proposed conditions whereby the continual compression of the solar atmosphere would constantly dissociate molecules at the same time that they were generating heat through their recombination. He had in effect created a model of the sun as a huge, externally self-perpetuating oxyhydrogen torch. It was in fact a perpetual motion machine in which dissociation accounted for the supply and storage of energy while combustion accounted for its release, but with the basic fuel remaining always the same.[9]

Williams' book appeared at the time when Darwin was feeling the pinch of Kelvin's time scale most severely. It is not surprising, therefore, to find Wallace enthusiastically recommending it as having solved "the great problem of the almost unlimited duration of the sun's heat in what appears to me a most satisfactory manner."[10] Wallace also mentioned that the book had impressed Lyell and the physicist William Grove. All three had apparently failed to grasp the fact that, at a very basic level, Williams' hypothesis, with its endless cycles of undiminished energy formations, violated the second law of thermodynamics, the very principle upon which Kelvin's opposition to uniformitarianism rested. This oversight is particularly significant since it seems to indicate that some of Kelvin's most important converts and opponents had failed to comprehend the fundamental premise underlying all of his arguments. Moreover, this same oversight appeared again as late as 1882 when C. William Siemens (1823-1883), a noted physicist and engineer and another of Kelvin's personal friends, published an hypothesis very similar to Williams' in the *Proceedings* of the Royal Society itself.[11] Thus it would appear that it was not only the geologists and biologists who missed the full implications of Kelvin's arguments. Physicists too suffered from the same disability, and they too sought occasionally to escape his restrictions.

Siemens and Williams notwithstanding, the most influential alternative to Kelvin's gravitational theory of solar heat was James Croll's stellar collision hypothesis. It was much more scientifically orthodox than either of the others, and although admittedly speculative, it did not violate any known physical principles. Furthermore, Croll, like Kelvin, kept his ideas alive by returning to them for defense or modification. First proposed in 1868 as a way of reconciling the physical estimates of the ages of the earth and sun, Croll never changed the substance of his solar hypothesis.[12] But stimulated by Kelvin's increasing dogmatism and more especially by Tait, he did change the grounds on which he justified his limited time scale. Croll was satisfied that 100 million years or even less would be ample to account for the earth's geological and biological history, and thus his original concern was to show that the physicists' results could be reconciled with geological data. He developed this same theme more fully in 1875 in *Climate and Time*, his most important work.[13] When Tait's *Recent Advances in Physical Science* appeared the next year proclaiming that the physical estimates of time must be reduced to fifteen or twenty million years, however, Croll felt obliged to reply. "It is the facts of denudation," he declared "which most forcibly impress the mind with a sense of immense duration, and show most convincingly the great antiquity of the earth."[14] No longer was his task the reconciling of geology with physics. Instead it was the *evidence of geology* which showed "with absolute certainty that [the earth's age] must be far greater than 20 million years."[15] This shift of emphasis was much more significant than the minor changes in Croll's hypothesis. He still believed that 100 million years was long enough for geology, but if subsequent discoveries indicated a need for a longer period, he saw no reason why his hypothesis could not account for double that amount or more.

As early as 1877 Croll had placed the physical arguments in a perspective that continued to evade most geologists and physicists for another two decades. The fallacy of the physicists' arguments was not their conviction that the sun's heat could not exceed its total store of energy. The mistake was in their failure to recognize that:

. . . it does not follow as a necessary consequence, as is generally supposed, that this store of energy must have been limited to the amount obtained from gravity in the condensation of the sun's mass. The utmost that any physicist is warranted in affirming is simply that it is impossible for him to *conceive* of any other source. His *inability*, however, to conceive of another source cannot be accepted as a proof that there is no other source. But the physical argument that the age of our earth must be limited by the amount of heat which could have been received from gravity is in reality based upon this assumption—that, because no other source can be conceived, there is no other source.

It is perfectly obvious, then, that this mere negative evidence against the possibility of the age of our habitable globe being more than 20 to 30 million years is of no weight whatever when pitted against the positive evidence here advanced that its age must be far greater.[16]

The burden of proof rested with the physicists, not the geologists. If geology could prove, as Croll thought it had, that the earth had existed with the sun to warm it for more than 20 million years, then the physicists' conclusions, and hence their assumptions and methods of reasoning, must be wrong.

Although Croll's solar hypothesis never attained the general acceptance of his theory of glaciation, it did offer hope to those who were disturbed by the physicists' time scale; and he continued to defend it, even extending it to include a general theory of nebular formation.[17] By the time of the appearance of *Climate and Cosmology* in 1886 and *Stellar Evolution* in 1890, Croll was as uncompromising as the physicists.[18] Finally Kelvin himself responded in 1887 and rejected the collision hypothesis as prohibitively improbable if not altogether impossible.[19] Croll's rejoinder, however, shows just how far he had come from the reconciliation of physics and geology. Reviewing once again the various calculations of the earth's age, he concluded that the "facts" of biology and geology were irreconcilable with Kelvin's explanation of solar heat. "I think it must now be perfectly evident," he exclaimed, ". . . that the mistake in regard to geological time has been committed by the physicist and not by the biologist."[20]

Croll also led the objections to Kelvin's theory of tidal retardation. As early as 1871 he argued that while the earth's velocity of rotation may indeed be decelerating, its present shape can provide no clue to its age. His argument had nothing to do with the possible plasticity of the earth's interior. Rather, he asserted that the denudation of the terrestrial surface at the equator would have effectively eroded away any bulge that might have existed at the time of its consolidation. Hence Kelvin's argument had no foundation. Again stimulated by Tait's assertions Croll repeated his argument several times during the next decade, but as this was the least restrictive of Kelvin's methods, he did not belabor the point.[21]

A much more telling rejection of the tidal retardation argument, and indeed of Kelvin's whole position on the age of the earth, came from George Darwin. At Kelvin's instigation Darwin had made the theory of tides his own. His whole theory of lunar genesis depended upon the principle of tidal retardation; and though he had relied principally upon the action of terrestrial rather than oceanic tides, he had originally seen no antagonism between his theory and Kelvin's geochronology. By 1886, however, he had become convinced that the mass of geological and biological evidence could no longer be subordinated to the increasingly restrictive time scales of Kelvin and Tait. And in the first of the series of addresses that during the next decade were to make the British Association a major forum for the renewed controversy over the earth's age, he reexamined Kelvin's arguments and publicly warned against the uncritical acceptance of his results.

The theme of Darwin's address was that scientists should recognize their relative ignorance of the great forces involved in terrestrial and solar activity and refrain from making dogmatic statements about the earth's age until more precise knowledge became available. His chief target was clearly Tait, and he passed quickly over the problem of solar heat to take up the two arguments on which Tait's most restrictive estimates had been based. He was also relatively mild in his treatment of the question of the earth's secular cooling. He was willing to accept its general validity so long as wide allowance for uncertainty—at least as great as those originally allowed by Kelvin—were retained. But he strongly denied any justification for Tait's 10 million years. He reserved his most critical remarks for the argument from tidal retardation where he accused Kelvin with failing to give a full account of the premises involved in the data he had used and thus with attributing to them far greater certainty than they could legitimately claim. Moreover, even if all of the data on the earth's rate of axial deceleration proved correct, he questioned whether the earth's present shape could give any clue to its age. Denudation and the probable plasticity of the terrestrial core might both work to obliterate any trace of its primordial bulge. Indeed Darwin felt that the whole argument was so uncertain that a limit of a billion years might be inadequate to allow a proper margin of error.[22]

Darwin was far from advocating a return to the uniformitarianism of the midcentury. Geological calculations of the order of John Phillips' 38 million to 96 million years still seemed to him significant, and he believed that Kelvin was fully justified in limiting the earth's age to "some such period of past time as 100,000,000 years." What he objected to was the high degree of precision that Tait and Kelvin attributed to their results. Too much remained to be learned to justify the making of dogmatic assertions. "At present," he said in conclusion, "our knowledge of a definite limit to geological time has so little precision that we should do wrong to summarily reject any theories which appear to demand longer periods of time than those which now appear allowable."[23] On the question of tidal retardation, Darwin certainly spoke with authority. His mathematics were as incontrovertible as Kelvin's had been, his assumptions as probable, and his results more compatible with geological evidence. His conclusions, therefore, deserved and got consideration. It is ironic that many of the references to his arguments neglected his assumptions and his allowances for uncertainty, and attributed far greater precision to his estimate of the moon's age than he himself felt that it deserved.

The possibility of the earth's having a viscous or fluid interior also had implications for Kelvin's analysis of the earth's secular cooling as well as for tidal retardation. Although generally conceded to be the most exact of the physical arguments, Kelvin's calculations concerning terrestrial cooling had been based upon conduction through a solid globe. If on the other hand, the earth's interior were fluid, as many if not most geologists believed, convection currents might appreciably alter his results. On these grounds Kelvin had a no more consistent critic than the Rev. Osmond Fisher (1817-1914), one of the last of the parson-naturalists and a pioneering geophysicist. Fisher's influential *Physics of the Earth's Crust*, first published in 1881, was devoted to demonstrating that a plastic substratum supporting a thin crust was essential to account for the condition of the earth's surface.[24] Fisher was particularly concerned with Kelvin's insistence upon a completely solid earth, but he also pressed the point that if this assumption proved false—as he believed it must—then the arguments limiting the earth's age would be invalid. Fisher backed up his contention with an impressive array of mathematics and calculations,[25] but at the same time he sharply criticized the physicists for ignoring geological evidence. In an open letter the following year, for example, he asserted:

> With respect to the yielding of the crust, I think we cannot but lament, that mathematical physicists seem to ignore the phenomena upon which our science founds its conclusions, and, instead of seeking for admissible hypotheses the outcome of which, when submitted to calculation, might agree with the facts of geology, they assume one which is suited to the exigencies of some powerful method of analysis, and having obtained their result, on the strength of it bid bewildered geologists to disbelieve the evidence of their senses.[26]

Fisher continued his opposition well into the nineties and sharply criticized Clarence King's calculation of time based upon the assumption of an entirely solid earth. He was immediately rebutted by two of King's supporters, Carl Barus and George F. Becker, but by 1893 the weight of geological opinion was clearly on his side.[27] Like many of his fellows, the age of the earth had not been Fisher's primary concern. He never bothered to make an independent calculation of time, but he recognized the importance of the problem for geology as a whole. Perhaps more significant, however, in his insistence that the hypotheses and results of the physicists be consistent with the evidence of geology, he represented the geologists' growing dissatisfaction with the high-handed attitude displayed by Kelvin and his followers.

Another persistent critic of the physicists' viewpoint was T. Mellard Reade. It has already been noted that Reade was a pioneering innovator in developing geological methods for determining the earth's age, and that he long held out for an estimate of time far beyond Kelvin's most liberal limits. He was also a defender of Lyellian uniformitarianism, and as early as 1876 had attempted to shift the burden of proof concerning estimates of geological time from the geologists to the physicists and mathematicians.[28] For Reade, geology was preeminently a practical, common-sense, inductive science. It was therefore "necessarily a safe science."[29] Physics and astronomy on the other hand were more exact and more mathematical, but also more deductive. Thus without detracting from the importance of the methods of astronomy and physics, Reade argued that they could not and should not displace the methods of geology in developing the geological history of the earth. His most important statement to this effect appeared in 1878, and it was little less than an all-out attack upon Kelvin's argument from the earth's secular cooling. Indeed, he denounced the whole argument as nothing more than "tremendous super-structure of inference" built upon unverifiable assumptions and unwarranted generalizations from inadequate data.[30] To support his contention Reade contrasted Kelvin's value of the earth's thermal gradient drawn from a few measurements around Edinburgh, with the wide range of gradients from different localities reported by the British Association's Committee on Underground Temperatures. He also dramatically illustrated how a small change in Kelvin's questionable data could cause a large change in his results, and how other equally small changes in his assumed conditions could yield results as varied as 25 million years and 1.6 billion years for the same calculation. Too many of Kelvin's data, including the earth's thermal gradient, its internal conductivity, and its supposed solidity, were mere assumptions; and even the unverifiable hypothesis that the earth was originally molten smacked of the cosmogonic reasoning that Lyell had exorcised from geology. Kelvin's method, in short, was completely alien to geology, and in Reade's opinion it was so uncertain that it could place no reliable limits upon the earth's age.

Reade was equally critical of what he considered to be hasty hypotheses and generalizations from geology. Thus he was as severe in his criticism of Wallace's calculation of time in *Island Life* as he was of Kelvin's physics.[31] In the end, however, it was the evidence of geology that had led him to oppose Kelvin, and it was that same evidence reinterpreted that led him finally to accept 100 million years as an adequate allowance for the earth's age. The fact that virtually all of the geological calculations leading to this shortened time scale had been influenced either directly or indirectly by Kelvin's results was conveniently overlooked.

By the time of Reade's conversion in 1893, the effects of Kelvin's early influence had come into direct conflict with his later conclusions. Even Geikie, who probably more than any other individual had spread the influence of Kelvin's original ideas among geologists, was repelled by the physicists' later estimates, by their arrogance, and by their utter disregard for geological evidence. His own belated calculation of time did not appear until 1892, but he took the opportunity afforded him as

president of the British Association to lay down a public challenge to
the physicists.

> After careful reflection on the subject, I affirm that the
> geological record furnishes a mass of evidence which
> no arguments drawn from other departments of na-
> ture can explain away, and which, it seems to me,
> cannot be satisfactorily interpreted save with an allow-
> ance of time much beyond the narrow limits which
> recent physical speculation would concede.[32]

Geikie had been one of the first to accept Kelvin's 100 million year
earth, and one of the most persistent in condemning the early uniformitarians for what he considered to be their excessive demands for
time. He still saw no reason to change that opinion. Indeed geological
calculations had confirmed it a dozen times over. But now as the
physicists continued inexorably to reduce their estimates of time, he
was convinced that somewhere they must be wrong. Geikie's intellectual roots were in midcentury geology, however, and though he had
helped to develop the quantitative methods which had raised geology
above mere description, he was not able to dispute Kelvin's arguments
upon their own ground. He could, nonetheless, confidently reject
them upon geological grounds.

> We must also remember that the geological record
> constitutes a voluminous body of evidence regarding
> the earth's history which cannot be ignored, and must
> be explained in accordance with ascertained natural
> laws. If the conclusions derived from the most careful
> study of this record cannot be reconciled with those
> drawn from physical considerations, it is surely not too
> much to ask that the latter should also be revised. . . .
> That there must be some flaw in the physical argument
> I can, for my own part, hardly doubt, though I do not
> pretend to be able to say where it is to be found. Some
> assumption, it seems to me has been made, or some
> consideration has been left out of sight, which will
> eventually be seen to vitiate the conclusions, and which
> when duly taken into account will allow time enough
> for any reasonable interpretation of the geological
> record.[33]

This opinion was shared by many other geologists as the century
entered its final decade. They were unable to refute the physical

arguments specifically, but they had grown increasingly uncomfort-
able with the restrictions imposed upon geology, and they were convinced that the physicists had somewhere made an error. For example,
Robert S. Woodward, the astronomer and chief geographer of the U.S.
Geological Survey, examined Kelvin's secular cooling argument and
declared it "very important if true." But then after dissenting from
three of the four assumptions upon which it was based, he concluded
that "although the hypothesis appears to be the best which can be
formulated at present, the odds are against its correctness."[34] Another
American, C. D. Walcott, struck much the same note when he declared:

> Of all the subjects of speculative geology few are more
> attractive or more uncertain of positive results than
> geologic time. The physicists have drawn the lines
> closer and closer until the geologist is told that he must
> bring his estimates of the age of the earth within a limit
> from ten to thirty million years. The geologist masses
> his evidence and replies that more time is required and
> suggests to the physicist that there must be an error
> somewhere in his data or the method of his
> treatment.[35]

Such skepticism was not confined merely to the physicists' later
results, however. While some geologists were impressed by the *relative*
agreement of the various chronological estimates, others were more
struck by the diversity of the *exact* numerical values. Thus the opinion
appears repeatedly, from physicists and geologists, Americans and
Englishmen—from George Darwin in 1886, Henry Shaler Williams in
1893, and Joseph Prestwich in 1895, to name but a few[36]—that no
accurate quantitative measure of time was possible given the gaps and
uncertainties in the data. These doubts even invaded those bastions of
certainty, the popular monographs. In *The Story of Our Planet*, for
example, T. G. Bonney, the professor of geology at University College
London, discussed the uncertainties involved in both the geological
and physical approaches to the problem of time and concluded that
neither could claim any degree of precision. He was willing to accept
from the physicists the conclusion that the earth is probably no older
than 100 million years and from the geologists that it must be older
than 30 million years, but beyond that he felt that geologists must stand
firm and not be swayed by the exaggerated claims of physics.[37]

By far the most radical statement regarding the uncertainties involved in determining the earth's age came from an American

geologist and anthropologist, William J. McGee (1853-1912). McGee's interest in time sprang in part from his concern with the apparent cyclic periodicity of various natural processes and in part from a desire to determine the date of the appearance of man. But it went far beyond either question, and led him to investigate the various ways in which the earth's age had been determined. Like Darwin and Bonney, he became convinced that all of the methods tried thus far, both physical and geological, contained far too many unknown variables to be granted any degree of final acceptance. Indeed, he concluded that the possibility of error was so great that the range of probable error normally allowed in physical and astronomical calculations was inadequate. He suggested instead that a "factor of safety" of the sort used by engineers—normally about six to ten times the estimated necessary strength of a structure—would be more appropriate for geological speculation. The geologist, however, needed an even greater factor of safety than the engineer since the further back in time he probed the more inexact his data became. McGee therefore proposed a factor of safety which went up exponentially with time. He suggested that a factor of 4 would be appropriate for the post-glacial time, a factor of 8 for post-Columbian, 256 for the Mesozoic era, and 1024 for the Paleozoic. In order to illustrate his argument, he made a calculation of time using an eclectic combination of previous geological methods. His resulting mean value for the earth's age was a startling 15 *billion* years, but because of the large factors of safety this result meant that the earth might be anywhere from 20 *million* to 15 *trillion* years old. Obviously, such a range rendered the disagreements between physics and geology superfluous.[38]

McGee's original skepticism applied equally to the results of both physics and geology. But the following year, in response to the speculations of Clarence King, his position hardened, and he took the physical arguments particularly to task. In April, 1893, the Geological Society of Washington held a symposium on the age of the earth which was devoted primarily to a discussion of King's speculations. There McGee repeated his previous conclusions with one significant modification. He still insisted upon a need for a factor of safety, and his method of calculating the earth's age remained much the same except for the correction of an arithmetical error which reduced his results to a mean of 6 billion years and limits of 10 million and 5 trillion years respectively. But the thrust of his criticisms now centered almost exclusively upon the physical estimates of time. While still admitting that great uncertainties existed in the geologists' data, he argued that their estimates were nonetheless based upon direct observation under actual conditions and that they dealt with real problems. They consequently eliminated many of the inaccuracies resulting from the complex and unknown factors involved. The physical arguments on the other hand were based upon idealized conditions and required simplifying assumptions which in McGee's opinion reduced them to interesting and instructive abstract problems, but which left them with very little resemblance to actual phenomena.[39] Thus, even though he still required large factors of safety for the geological data, he joined the ever growing band of geologists who were willing to pit the results of geological investigation against the hypotheses of the physicists.

McGee's position was radical by any standard, but it was actually little more than an exaggerated extension of the doubts and discontent felt by many of his contemporaries. The high-handed pronouncements of Kelvin and Tait, and the excessive restrictions that they and King would place upon the limits of time had greatly accelerated the tempo of the geologists' discontent. Proud of their new quantitative methodologies and confident of the adequacy of their data, the geologists were no longer willing to be cowed by the pronouncements of physics. Few of them as yet could refute the physicists in their own terms, but like Geikie they were certain that an error had been made. Moreover, they were not alone. As Darwin's address made clear, the physicists were themselves beginning to question the dogmatism of the older generation. It was in this atmosphere of doubt and discontent that Salisbury unknowingly turned Kelvin's results against the accumulated weight of geological and biological evidence. The response was immediate, and it set a pattern of controversy that was to continue well into the new century.

The Challenge

The man who finally challenged Kelvin directly upon his own ground was John Perry (1850-1920), a former assistant and occasional collaborator of Kelvin's, an accomplished mathematician and engineer, and a man noted for diligence, ingenuity and heterodox opinions. Perry had originally been reluctant to criticize Kelvin's views, for although he attached little weight to the arguments from tidal retardation and solar heat, he had expressed the opinion in 1890 that the assumptions, data, and mathematics of the secular cooling argument were perfectly sound.[40] Furthermore, he confessed to being repelled by the vagueness of the conditions which had to be assumed in any quantitative geological problem. He had consequently evaded the requests of several friends that he examine Kelvin's calculations. Salisbury's address changed his mind. Kelvin's results had been used unequivocally to discredit the direct evidence of the best authorities in geology and biology, and Perry now felt that it was his "duty to question Lord Kelvin's conditions."[41] Perry had not actually heard Salisbury deliver his address, but upon reading it he began work immediately. Within weeks he had drawn up what amounted to a brief against Kelvin's assumptions, had circulated it privately among a number of eminent scientists, and had sent it as an open letter to *Nature*, where it appeared in January, 1895.[42]

Having once agreed to question Kelvin's theories, Perry went immediately to the heart of the matter. For thirty years Kelvin's calculations had been repeated by schoolboys as an abstract mathematical exercise. And, during the same period the general summaries of his arguments and conclusions had appeared countless times. But in spite of occasional questions, the details of Kelvin's framework of preconceptions and assumptions, especially those simplifying assumptions made to facilitate calculation, had been lost from sight. It was upon those that Perry focused attention. Aided by the calculations of Oliver Heaviside, he showed that the problem of a cooling earth would give very different results if certain of Kelvin's simplifying assumptions were omitted. If contrary to Kelvin's assumption the earth's conductivity were not homogeneous but were assumed to be greater near the center than at the surface, then Kelvin's estimate of the earth's age would have to be increased, perhaps dramatically. As an example Perry found that assuming an internal conductivity only ten times that of the crust would increase Kelvin's estimate of time by a factor of fifty-six. He was careful not to say that the earth's internal conductivity is greater than that at the surface; he merely pointed out that the differences in conductivity at various points on the surface itself made it conceivable that differences might exist in the interior. He did insist, however, that some degree of fluidity must exist in the earth's core, and that its thermal conductivity must therefore be supplemented by convection. On the whole, Perry believed that an increased internal conductivity was highly probable, but the real importance of his argument was the demonstration that, within the limits of scientific knowledge, other conditions than those assumed by Kelvin were possible, and that consequently the question of the earth's age was not definitely closed.[43]

In general, the pre-publication response to Perry's position seems to have been favorable, but his appeal to Tait for advice on how to convince Kelvin drew a typically arrogant response. The ensuing correspondence, some of which Perry published with his original argument, served to illustrate the inflexible stand taken by Tait and Kelvin and to intensify Perry's opposition. Tait began by dismissing the problem out of hand. He admitted that Perry's calculations would be correct if his assumed conditions could be proven. But upon what grounds, he asked, could one judge between the plausibility of Perry's conditions and Kelvin's since no data whatever was available to test either case? What then was Perry's point? And why bring mathematics into it at all? Perry's response was to herald a new approach to Kelvin's pronouncements. "Surely," he asserted, "Lord Kelvin's case is lost, as soon as one shows that there are *possible* conditions as to the internal state of the earth which will give many times the age which is your and his limit."[44]

Perry also took the opportunity to examine the arguments from solar heat and tidal retardation which always before he had regarded as secondary. Again he offered an alternative set of assumptions, and asserting that they were not only possible but probable, charged that it was for Tait and Kelvin "to prove the negative." Once again, still more abruptly, Tait refused to assume the burden of proof. The debate continued for several weeks with Perry giving the reasons for his assumptions but steadfastly insisting that Kelvin and Tait must assume the burden of proof for theirs. The real question, he maintained, was: "Had Lord Kelvin a right to fix 10^8 years, or even 4×10^8 years, as the greatest possible age of the earth?"[45]

Finally, Kelvin himself responded, and though his tone was more moderate than Tait's—he was, after all, writing to a successful former student—his convictions were no more flexible. After expressing his doubts that the earth's internal conductivity could be anywhere near as high as Perry assumed it might be, he conceded that on the basis of underground heat alone the upper limit of the earth's age might possibly be set at 4000 million rather than 400 million years. But the sun's heat, he insisted, still limited the earth's age to a few score million years. His concluding remarks certainly left no doubt of his position:

> So far as underground heat alone is concerned, you
> are quite right that my estimate was 100 millions, and
> please remark that that is all Geikie wants; but I should
> be exceedingly frightened to meet him now with only

20 million in my mouth.[46]

As Perry had hoped, Kelvin did reexamine the secular cooling argument, but with the result that he completely rejected Perry's proposed conditions and instead came out in favor of the hypothesis and results of Clarence King.[47] Thus, rather than converting Kelvin to a more liberal view of geological time, Perry had succeeded in prodding him into a firmer acceptance of the more restrictive arguments.

In his final rebuttal Perry paid particular attention to the question of solar heat, and again he presented an array of possible alternatives to Kelvin's assumptions. The thrust of his argument was unchanged. He did not contend that his assumptions were true; he asserted only that so long as they were possible, Kelvin must either prove his own assumptions or forego placing absolute limits on the earth's age. Perry also devoted more attention to the possibility of fluidity in the earth's interior and concluded that it utterly negated King's position. As for the geological arguments, he believed them to be conclusive in demanding more time than either Kelvin or King would allow. Most geologists, he conceded, would be satisfied with 100 million years and some with quite a bit less, but unless physics could *prove* otherwise, he saw no reason to bar the possibility of much greater limits. The weight of direct evidence was on the side of biology and geology; the burden of proof therefore must be assumed by physics.[48]

The reactions to Perry's position were immediate and generally favorable. Even before his final rebuttal, the pages of *Nature* were carrying letters from geologists who welcomed his proposals as a possible means of escape from the impasse existing between physics and geology. Geikie, for example, recognized that Perry's assumptions remained to be established, but believed that they offered a definite alternative to Kelvin's hypotheses. He no doubt felt a sense of vindication as he wrote: "But there seems at present every prospect that the physicists will concede not merely the 100 millions of years with which the geologists would be quite content, but a very much greater extent of time."[49] William J. Sollas (1849-1936), on the other hand, viewed the exchange as an invitation for geologists to get down to work. Personally committed to a relatively short sedimentary time scale, Sollas, the professor of geology at Oxford, professed a ready willingness to trade a few million years for a hypothesis which would account for fluid shells in the earth's interior. But if the new physical arguments allowed more time, it could readily be credited to the virtually unknown Precambrian. In the meanwhile, "Since physicists do not seem to be in complete accord on the question of the time which has elapsed since the earth first permanently crusted over, it may perhaps be as well to investigate the evidence to be obtained from stratified deposits."[50] A somewhat similar approach was taken by Osmond Fisher. Although willing to accept Kelvin's contention that Perry's assumptions were unsupported by the available evidence, Fisher took the opportunity to reassert his pet argument for the earth's internal fluidity, and concluded that convection currents would have essentially the same effect as Perry's increased conductivity in extending the earth's age.[51]

Perhaps the most enthusiastic support, however, came from the side of biology. For years biologists had taken little part in the debate over the earth's age. Despite occasional remarks like Ernest Haeckel's: "We have not a single rational ground for conceiving the time requisite [for evolution] to be limited in any way,"[52] most biologists had taken their clue from Huxley and had let biology take its time from geology—or if necessary from physics. Huxley himself had repeated this opinion several times and only a few months before Salisbury's speech, had written:

> I will not presume to question anything, that on such ripe consideration, Lord Kelvin has to say upon the physical problems involved [in determining the earth's age] . . . . And I take the opportunity of repeating the opinion, that whether what we call geological time has the lower limit assigned it by Lord Kelvin, or the higher limit assigned by other philosophers . . . the problem of the origin of those successive Faunae and Florae of the earth, the existence of which is fully demonstrated by paleontology remains exactly where it was.[53]

In their separate ways both Darwin and Wallace had accommodated themselves to Huxley's doctrine, but as the physicists' allowance for time grew progressively more restrictive, several younger biologists grew restless. One of these, Edward B. Poulton (1856-1943), the professor of zoology at Oxford, had been especially provoked by Salisbury's speech and was one of those who appealed to Perry to make a reply. He was further provoked by the arrogance of Tait's subsequent exchange with Perry, and when the opportunity came, he rose to the defense.

Once again the annual meeting of the British Association served as a forum for the controversy. Speaking as president of the zoological section in 1896, Poulton took limited issue with Huxley's dictum on time. He agreed that the biologists could follow the geologists in determining the time covered by the sedimentary record. But he argued that biology must take the lead in establishing the time necessary for the progress of evolution before the fossil record began. Speaking as a biologist, his intent was not to extend the geologists' estimate of the age of stratified rocks, but to extend the accepted limits of the earth's age prior to the formation of existing strata. This aim, however, required that he first refute the arguments of the physicists. His method of approach was threefold: he used Perry to demonstrate the uncertainties in the physical hypotheses; Geikie to illustrate the inconsistencies between the results of physics and geology; and his own work to present the evidence by which biology justified its demands for time.[54]

Poulton's account of Perry's position was quite perceptive. He not only presented the details of the alternatives to Kelvin's assumptions, but after consultation with Perry, elaborated at length upon their implications. Throughout the discussion his emphasis, like Perry's, was upon the uncertainty of Kelvin's results and upon the necessity for the physicists to assume the burden of proof for their assumptions. He made it clear, moreover, that contrary to the kind of dogmatic assertions made by Tait, neither he nor Perry had proposed a new chronological limit. They had merely shown that one was possible. His treatment of the geological argument was less satisfactory. In choosing to rely upon Geikie's British Association address of 1892, he had chosen one of the simplest geological calculations of time available and one that concealed simplifying assumptions that were as questionable as those of the physicists themselves. Moreover, in presenting Geikie's numerical data, he greatly overstressed the precision of the results obtained. And then by casually reducing the geologists' wide range of possible minima to a simple average, he attributed still more precision to a result based at best upon rather loose assumptions. In fact his bold assertion that 400 million years was the minimum time which geological evidence would allow for the formation of existing strata was totally inconsistent with Geikie's own conclusions.

In turning finally to biology, Poulton was again on firmer ground. Here much of his argument was devoted to presenting specific cases to illustrate that even the earliest fossils show a high degree of specialization and development. Then citing several semi-empirical principles which he considered irrefutable—that evolution results first in the divergence of general characteristics and then of specific characteristics, that natural selection requires much longer to alter simple organisms than more complex ones, and that the origin of no phylum can be found anywhere in the stratified record—he proceeded to argue that the degree of specialization found in the lower fossils can only be accounted for by a very long period of evolution prior to the beginning of geological record. This, of course, was exactly the position that Darwin had taken. But now instead of yielding to the demands of physics, Poulton presented his evidence as proof that the physical arguments must be wrong. No longer was he willing for biology to adjust its conclusions to the dictates of geology or physics. In its own sphere its results must stand equal or ahead of those of its sister sciences, and whatever physics might indicate, "Natural Selection will never be stifled in the Procrustean bed of insufficient geological time."[55]

Once stimulated by the new challenge, Kelvin returned to the question of geological time with a vengeance, and in June 1897 he delivered his last major address on the subject.[56] It was a wide-ranging pronouncement beginning with a renewed attack upon the uniformitarianism of Lyell and Darwin, proceeding through his old arguments with detailed reasons for the subsequent reductions in his original estimates, and concluding finally with new speculations about the origin of continents and oceans and the conditions necessary for the commencement of life. His views on his earlier arguments remained essentially unchanged: the argument from tidal retardation was based upon sound mechanical principles but was the least exact in its results, while the argument from solar heat was the most restrictive. The bulk of the address, however, dealt with additional comments on terrestrial heat and the consolidation of the earth's crust. Thanks to the work of King and Barus, Kelvin now believed that the arguments from solar and terrestrial heat had been reconciled. Perry's argument could be disregarded entirely. If anything Barus's experiments indicated a diminution of the earth's thermal conductivity with depth rather than the augmentation postulated by Perry. Indeed, Barus's data on the thermal properties of rocks, the same data used by King, seem to have been greatly responsible for the assurance with which Kelvin spoke of "half an hour after solidification," or "the temperature a few years after solidification." And it was this tone of assurance which set the argument apart, rather than any new information that it contained. Gone were the allowances for uncertainty. Kelvin was content to accept King's 24 million years as the best estimate available until some future discovery allowed it to be lowered still more. If by chance the data should push the date of consolidation further back in time, then the solid earth would wait for a while until the sun was hot enough to warm it.

Kelvin's address was reprinted several times, and in 1899 it appeared in the American journal, *Science*, where it was read by Thomas Chrowder Chamberlin (1843-1928), the professor of geology at the University of Chicago. For years an outstanding glacial geologist, Chamberlin had only recently turned to the cosmogonal speculations which led him to question Kelvin's position. But immediately upon reading the address, he began a detailed reply. His theme, like Perry's, was the air of certainty which Kelvin gave to speculations that could by no means be considered certain. Courteously but firmly he suggested that the use of such phrases as "certain truth," "very sure assumptions," and "half an hour after solidification" were so misleading as to convey untruth. "Assumption" rather than "very sure" should have been the key word in the physicist's statements, he asserted.[58] Such concern was not new to Chamberlin. Some ten years earlier he had warned of the dangers of rigid adherence to a single hypothesis. Without the beneficial check of alternative working hypotheses, he believed that premature explanations too often evolved into ruling theories while impartial intellectual inquiry deteriorated into the search for facts to support the theories.[59] It was just this danger that he saw in Kelvin's air of emphatic certainty. Chamberlin also had another reason for questioning Kelvin, however. In collaboration with Forest Ray Moulton (1872-1952), a young astrophysicist at Chicago, he had begun to develop an alternative to Laplace's long established nebular hypothesis, an alternative which, should it prove valid, would undermine all of Kelvin's arguments.[60]

In 1899, what was to become the planetesimal hypothesis was far from complete. Moulton and Chamberlin had developed their reasons for questioning the nebular theory, but had not yet worked out the details of their alternate hypothesis. In rebutting Kelvin, therefore, Chamberlin had to content himself with taking up each of the older man's arguments, identifying the underlying assumptions, and presenting alternative assumptions which might invalidate them.[61] The foci of the attack were Kelvin's unquestioning acceptance of the nebular theory and his subsequent assumption of a molten primordial earth. The argument against the nebular theory was due primarily to Moulton and was based upon recent developments in the kinetic theory of gases. According to Laplace, the sun and solar system had been formed by the consolidation of a gaseous nebula acting under the influence of gravitation. Moulton's calculations, however, indicated that in a gaseous state, the molecules of the earth's constituent materials would have velocities that were far too great to be overcome by weak gravitational forces alone, and thus consolidation could never take place. The alternative was to assume an aggregation of much larger, and hence cooler and slower particles, in other words, a confluence of meteors. This, however, was just the assumption that Kelvin had made. After adopting Laplace's hypothesis as an initial principle, he had based his calculations upon the dynamics of a sudden influx of meteoric matter. Chamberlin's disagreement lay in the assumed suddenness of the influx. There was no reason, he argued, why the earth could not have formed slowly through the gradual accumulation of meteoric material, so slowly, perhaps, that it may never have been molten at all. Chamberlin believed, in fact, that the assumption of an initial molten earth was due almost entirely to the uncritical acceptance of the nebular hypothesis, and he questioned whether Kelvin or anyone else had ever actually calculated how rapidly the postulated influx of meteors would take place.

In contrast to what seemed to him the speculative nature of the molten-earth hypothesis, Chamberlin believed that there was definite geological evidence to support the assumption of a primordial solid earth. There were also other decided advantages. In the first place, the heterogeneous nature of the earth's crust seemed to him far more consistent with a slow heterogeneous accumulation of matter than with an initial homogeneous molten mass. Secondly, if one assumed some degree of internal terrestrial plasticity, as many geologists did, then it would follow that the heterogeneous materials would gradually adjust their positions to attain a maximum density. They would thus slowly convert a part of their gravitational potential energy into heat. This assumption, which was directly analogous to Helmholtz's theory of solar heat, would account for the earth's high internal temperature through compression while providing for a much slower dissipation of the available energy than Kelvin's hypothesis would allow. It would therefore delay the establishment of Kelvin's measured thermal gradient and extend his estimate of the earth's age. Moreover, the continued slow influx of meteors would itself provide energy to help offset the heat lost by radiation and extend the permissible limits of time still further. Finally, Chamberlin believed that this postulated slow internal compression and stratification provided a more satisfactory mechanism for mountain formation than the common hypothesis of thermal contraction.

At another level, Chamberlin argued that a solid earth, increasing gradually in volume, would probably be able to sustain an atmosphere and support life when only two-fifths its present size. Thus life and the process of evolution might have begun long before the present terrestrial crust was formed, while the continued influx of meteors might in itself have maintained a temperature sufficient to sustain life even in the absence of adequate solar heat. Under such circumstances Kelvin's second argument would also lose its force. Chamberlin was careful to emphasize that he did not consider his hypothesis substantiated, but neither was Kelvin's hypothesis of a molten earth. Both still had to pass the tests of criticism and proof. It was up to the astronomers and physicists, therefore, to prove that the aggregation of meteors would occur rapidly enough to form a white hot liquid earth, and until then, their assumptions were as much on trial as the chronological speculations of geology and biology.

Chamberlin's replies to Kelvin's other arguments were less well-developed than his explanation of the earth's formation. The argument from tidal retardation depended, after all, upon the earth having solidified from a molten state. Thus, having rejected its major premise at the outset, Chamberlin contented himself with criticizing the oversights in Kelvin's hypothesis and with attacking Darwin's lunar theory which had been used to support it. By contrast, however, his discussion of solar heat was extremely speculative, even to the point of reintroducing the nebular conditions that most of his paper had attempted to refute. Was it not possible, he suggested, that in the past a more diffuse sun filling much of the earth's orbit might have supplied enough heat to support terrestrial life while losing far less total heat than it now does? Or more speculatively, was it not possible that nebular conditions beyond the earth's orbit may have conserved the limited supply of solar energy by somehow reflecting it back into the solar system? His best remembered speculation, however, was a confession of ignorance about the nature of the sun itself:

> Is present knowledge relative to the behavior of matter
> under such extraordinary conditions as obtain in the
> interior of the sun sufficiently exhaustive to warrant
> the assertion that no unrecognized sources of heat
> reside there? What the internal constitution of the
> atoms may be is yet open to question. It is not improb-

able that they are complex organizations and seats of enormous energies. Certainly no careful chemist would affirm either that the atoms are really elementary or that there may not be locked up in them energies of the first order of magnitude. No cautious chemist would probably venture to assert that the component atomecules, to use a convenient phrase, may not have energies of rotation, revolution, position, and be otherwise comparable in kind and proportion to those of the planetary system. Nor would they probably be prepared to affirm or deny that the extraordinary conditions which reside at the center of the sun may not set free a portion of this energy.[62]

Leaving aside the appearance of prophecy, this statement epitomizes much of the spirit of Kelvin's challengers. As Croll had pointed out years before, the universe is not limited by the scientist's conception of it. The inability to conceive of a source of solar energy other than gravity was not proof that other sources did not exist. In its harshest terms, Perry and Chamberlin had asserted that ignorance alone could not justify Kelvin's claims of certainty.

It was not necessary to build a new cosmogony or even to postulate extraordinary solar or nebular conditions in order to challenge Kelvin's geochronology, however. The possibility of rival assumptions was in itself enough. Time had wrapped a cover around Kelvin's mathematical mill, which Perry, Chamberlin, and others had ripped off. But even when fully revealed to criticism, many of Kelvin's assumptions still held firm. The nebular hypothesis, for example, was no more speculative than the planetesimal hypothesis and in some respects still had the advantage of simplicity. And the laws of thermodynamics and Newton's mechanics still stood untarnished. But the paucity of Kelvin's evidence and his superstructure of assumption had been laid bare, and so long as those assumptions remained unproven, the responsibility for defending his claims to certainty now rested with him.

The Loyal Opposition

Perry had stripped the facade of mathematical certainty from the physicists' arguments and revealed the framework of hypothesis and assumption underneath. Perhaps his attack had not been necessary.

The mass of evidence accumulated by geology and paleontology had itself convinced many geologists that the physicists must be wrong. Perry, however, had shown just where they might be questioned, and in so doing had helped to nurture the geologists' growing feeling of confidence. He had not offered an alternative chronology of his own, but he had at last driven home Huxley's warning about the limitations of the mathematical mill. After three decades the wraith of mathematical certainty no longer haunted geology.

Once shown the fallibility of the physicists' assumptions, the geologists could hardly ignore the uncertainties inherent in their own methods. Thus, although no one seriously considered returning to the vagueness of the midcentury uniformitarians' indefinite eons, a recurrent theme was the inadequacy of any method to measure the age of the earth accurately in years. Estimates of time were made—indeed the nineties were particularly productive compared to the relative indifference of the eighties—but the conventional methods no longer gave such homogeneous results. Released from the restrictions of the physicists, the geologists felt free to disagree among themselves as well. A few of them consequently still supported Kelvin's 100 million-year earth or even one of his more restrictive estimates. Others would accept no less than several hundred million years. Many, perhaps most, however, still followed Geikie's lead. They were content with 100 million years on the basis of geological calculations, but they left open the possibility of much longer estimates should subsequent evidence demand it. Aware of the shortcomings in both the physical and the geological data, the geologists had grown ever more confident of their methods; and their confidence showed clearly in their willingness to question and doubt.

Geikie remained the spokesman for the moderate view. His speech in 1892 had been a cry of frustration at the excesses of the physicists, excesses as blatant as those of the early uniformitarians. A few years later he again summed up his feelings with customary eloquence:

> Geologists have not been slow to admit that they were in error in assuming that they had an eternity of past time for the evolution of the earth's history. They have frankly acknowledged the validity of the physical arguments which go to place more or less definite limits to the antiquity of the earth. They were, on the whole, disposed to acquiesce in the allowance of 100 millions of years granted them by Lord Kelvin, for the transac-

tion of the whole of the long cycles of geological history. But the physicists have been insatiable and inexorable. As remorseless as Lear's daughters, they have cut down their grant of years by successive slices, until some of them have brought the number to something less than ten millions.

In vain have geologists protested that there must somewhere be a flaw in a line of argument which tends to results so entirely at variance with the strong evidence for a higher antiquity, furnished not only by the geological record, but by the existing races of plants and animals. They have insisted that this evidence is not mere theory or imagination, but is drawn from a multitude of facts which become hopelessly unintelligible unless sufficient time is admitted for the evolution of geological history. They have not been able to disprove the arguments of the physicists, but they have contended that the physicists have simply ignored the geological arguments as of no account in the discussion.[63]

As Geikie wrote, Perry was pointing out where the flaw in the physical arguments might be found. But for Geikie and many others, Kelvin still spoke for physics, and Perry's challenge had only pushed him toward the still more restrictive conclusions of Clarence King.

Geikie's response to Kelvin's final pronouncement is quite revealing. Speaking again before the British Association in 1899, his frustration with the physicists' high-handed attitude was apparent, and his own words were much more authoritative and polemical than ever before. In a measured crescendo, he reviewed the history of the question of geological time, citing the excesses of the early uniformitarians, the impact of Kelvin's hypotheses, the subsequent reductions in the physical estimates, the resulting controversies, and finally the growing discontent of the geologists. Then in what amounted to a declaration of independence from the autocratic pronouncements of Kelvin and Tait, he followed up his defense of geology and biology with a challenge to geologists to put their arguments on a sound foundation of measurement and experiment. There were no new calculations in Geikie's statement. Like Perry and Chamberlin, he had concluded that too little was known to make meaningful calculations. It was time, instead, to gather evidence unfettered by the speculations of the past.[64]

Two points in Geikie's address deserve particular attention: his repeated calls for more geological measurements and his defense of the concept of *uniformity* as distinct from the *uniformitarianism* of Lyell and his followers.[65] In the latter case he argued that nowhere in the geological record is there any evidence to support the physicists' contention that geological processes were more violent in the past than in the present. On the contrary, every evidence seemed to show that the rate of geological activity and the conditions for the maintenance of life have been remarkably constant since the formation of the oldest sedimentary rocks. Moreover, he continued, this conclusion was based upon evidence available at every hand whereas the physicists' arguments, for all of their assurance and mathematical formulae, were founded upon mere assumptions which even they—here he mentioned Perry and Darwin specifically—were no longer willing to accept without question. Geikie conceded here that the geological evidence also lacked precision, and hence his call for the continued development of techniques that would put geology on a sound experimental footing. Only by "accurate, systematic, and prolonged measurement," he declared, could geologists accomplish the essential task of eliminating "the predominant vagueness" found in too many geological works.[66]

As his historical survey reveals, Geikie was still committed to a relatively short time scale, and so his attack was leveled primarily at the dogmatism and excessive restrictions of the later physical estimates. He still praised Kelvin for rescuing geology from the vague conceptions of limitless time, and by exaggerating the "recklessness" of the chronological demands of the midcentury uniformitarians, did much to perpetuate the myth that they were basically unscientific.[67] On one point, however, he was adamant. For too long geology had yielded meekly to the demands of physics while the physicists ignored the geologists' arguments. The data of geology and paleontology might lack quantitative precision, but it could not be ignored. Indeed, the time had come for geologists to attack "the stronghold of speculation and assumption in which our physical friends have ensconced themselves."[68] Asserting that infinitely more was known of the earth's history than of the cosmic forces postulated by the physicists, Geikie too joined the chorus of those demanding that the physicists assume the burden of proof for their speculations. He still believed that 100 million years would prove adequate for the needs of geology, but if the discoveries of paleontology demanded more, he saw no reason to deny whatever time they required. "Until, therefore, it can be shown that geologists and paleontologists have misinterpreted their records, they are surely well within their logical rights in claiming as much time for the history of this earth as the vast body of evidence accumulated by them demands."[69] Kelvin was certainly right to fear meeting Geikie with only twenty million years in his mouth.

Others were not so reluctant as Geikie to use the quantitative data already provided by geology. Some three years earlier, John G. Goodchild, the curator of the Geological Survey Collections at the Edinburgh Museum of Science and Art, had thoroughly re-examined the geological data with interesting results. Goodchild's approach was eclectic. Borrowing freely from the methods pioneered by Phillips, Reade, Geikie, and others, he determined the duration of each separate geological period using what he regarded as the most appropriate and accurate method for each. Thus he determined the duration of the Miocene epoch from the rate of erosion in valleys, the length of the Tertiary and Cretacious periods from the formation of marine limetone, and the time represented by unconformities from the evidence of adjacent denudation. Going beyond the measurements of the strata, he cited the evidence of biology and paleontology to argue for a long period of Precambrian development; and where necessary, he used the change in fossil forms to justify assigning long periods for the formation of comparatively thin strata. In all he determined the ages of fifteen geological periods since the beginning of the Cambrian, totaling just over 700 million years, with at least an equal duration required for the Precambrian.[70]

It is the tone of Goodchild's argument rather than his numerical results that is most revealing, however. He considered the number itself of little importance. Indeed, given the uncertainties in the data available, he considered it impossible to determine the earth's age in years or even millions of years. "All we can feel sure of," he declared, "is that the records of the rocks fully justify us in claiming for the earth an antiquity so vast as to be far beyond the power of the human intellect to grasp."[71] But of that much geology could be certain. Thus in spite of his frequent repetition of the phrase "no reliable data," Goodchild shows an unshakable confidence that geology was on the right course. His quantitative results might be questionable and his data largely speculative, but as Perry and Croll had already demonstrated, so were those of the physicists. The real issue, therefore, was not the precision of his results, but the overwhelming weight of his evidence:

  . . . if it can be shown, on sufficiently good grounds

> that the changes that have taken place in a period so
> recent as the Tertiary Period require for their accom-
> plishment a measure of time so much in excess of that
> which, on mathematical grounds we are told is all that
> can be allowed for the time since it was first possible for
> life to exist upon the earth, then, I would respectfully
> submit, there must be some serious error in the data
> upon which those computations are founded. I ven-
> ture to think that such must be the case . . . . In the
> meantime, geologists will probably agree with me in
> thinking that we may disregard the view [of the
> physicists] regarding the Age of the Earth.[72]

Goodchild believed he had shown that and more. If the question of
proof remained, it rested with physics, not geology.

Perhaps the most remarkable thing about this display of confidence
was that it occurred without the stimulus of any real advance in the
relevant geological evidence. The store of specific information had
increased tremendously during the preceding decades, but the accepted estimates of the rates of geological processes, most of which
were still based upon loose averages and incomplete data, had not
appreciably changed. Goodchild had frankly borrowed much from his
predecessors, but in attitude he was far removed from the geologists
who had pointed with relief and pride to the agreement between their
results and those of the physicists. His 700 million years since the
beginning of the Cambrian would seem extravagant even to geologists,
he conceded, and since his data was frankly based as much on speculation as on evidence, he fully expected to be proven wrong in some
particulars. Significantly, however, he concluded that when proven
wrong, he was "quite prepared to find that, as our knowledge of
Geology and Biology advances, we shall feel it necessary to extend
rather than abbreviate the estimate I have put forward."[73]

If conventional methods could occasionally lead to radical results,
novel methods could give conventional results. In the spring of 1899
John Joly (1857-1933), the professor of geology and mineralogy in the
University of Dublin, announced the first entirely new approach to the
earth's age to appear in several decades, a calculation of the age of the
oceans based upon their sodium content.[74] But although the approach
was new and the selection of data different from those used in any
previous calculation, Joly's result turned out to be in excellent agreement with the dominant geological estimates. It was, moreover, so

packed with numbers and calculations that it became one of the most powerful arguments in the opposition to the physicists' restrictive scale.

The basic postulates of Joly's method were remarkably simple. If it could be assumed that the oceans were initially precipitated salt-free from the primordial atmosphere, and that their store of sodium had been supplied at an approximately uniform rate by the rivers of the world, and that the sodium has remained trapped in solution in the oceans since the process began, then the simple ratio of the sodium now contained in the oceans divided by the rate of supply from river sediment, should give the time required for its accumulation, or in other words, the approximate time since the earth's surface cooled below 212°F. The data for such a calculation already existed. Analyses of the mineral content of the oceans and major rivers had been made, as had an approximate estimate of the volume discharge of the world's rivers and several estimates of the total volume of the oceans. From these data it was a simple calculation for Joly to determine that as a first approximation his assumptions would yield a minimum of 90 million and a maximum of 99.4 million years for the oceans' age, depending upon which estimate of oceanic volume he accepted. It was an interesting result, but far too simplistic to be satisfactory to its author. The major part of his paper, therefore, was devoted to the identification and evaluation of any perturbations which might either vitiate or further refine his conclusions.

One question of course was paramount, the validity of assuming uniform rates of past geological processes. Like Geikie, Joly was more than willing to attack the extremes to which he believed uniformitarianism had been pushed by Lyell's followers. But he also believed that the principle of uniformity "rightly defined" was fundamental to geology. By rightly defined he meant simply "that we may justly prolong into the past and future the activities of today until sufficient reason be shown to interrupt them by catastrophe or change."[75] This definition of uniformity did not require that processes not change with time, nor that the rates of geological activity be considered absolutely constant. It required only that an approximate balance of geological activity be assumed in the past. It was for this reason that he chose sodium, a product of solvent denudation, as his measure of time. He believed that he could show that the secular changes in the earth's surface which might alter the rate of solvent denudation would quantitatively annul each other, and thus he could eliminate most of the uncertainties implicit in assuming simple uniformity. Some uncertainty would necessarily remain, he conceded, but after carefully arguing his case for a uniform average rate of solvent denudation he concluded that:

> There appears no good reason to suspect that our broad uniformitarian principles are leading us into considerable error where, more especially, such disturbing causes as we are compelled to recognize are both of positive and negative signs. But the whole consideration should undoubtedly lead us to widen the margin we allow for error in our estimate of geological time.[76]

Uniformity was only one of the possible objections to be answered, however, and Joly seemed intent upon overwhelming his readers with the variety and detail of his evidence, calculations, and arguments. Trained in physics and engineering as well as geology, he had also made original contributions to chemistry and astrophysics, and he drew upon them all in his argument. Beginning with the generally accepted primordial fireball, he explained how the balance of pressure and temperature would lead to a comparatively sudden precipitation of the oceans soon after the earth's crust hardened. He then began to enumerate possible changes in terrestrial conditions which might alter his calculations. An excess of hydrogen chloride in the primitive atmosphere, for example, would probably accelerate the initial rate of solvent denudation, thus requiring a reduction in his tentative estimate of time. A similar reduction might also be required to account for some sodium being absorbed directly from the minerals covered by the primordial sea. On the other hand, if an allowance were made for some salt being recirculated by winds and sediment, the initial estimate would have to be extended. In all, Joly isolated seven factors which might tend to reduce his estimate of time, and seven which might increase it; and he attempted to deal with each in detail. Although much of his data was extremely scanty—some of them indeed were little more than speculations—he tried, insofar as possible, to supply quantitative corrections based on the best information available, and he made every effort to allow for possible errors. Thus for geologists sensitized to the need for quantification, his array of numbers, measurements, and calculations was impressive indeed. And most of them, no doubt, were quite content with his conclusion that the oceans were probably between 80 million and 90 million years old.

Nonetheless, the immediate reactions to Joly's conclusions were decidedly mixed. Sollas, for one, accepted the method wholeheartedly, but argued for a much greater rate of supply of sodium in the past than at present, and hence a corresponding reduction in the overall estimate of time.[77] William Mackie, on the other hand, provided a point by point discussion of a half dozen possible errors in Joly's assumptions and threw the whole problem out as fundamentally insoluble.[78] The most perceptive review, however, came from Osmond Fisher, who pointed out that the sodium content of sediments as early as the Silurian was little different from that of the most recent formations, and concluded in consequence that the salinity of the oceans had probably not changed much since the beginning of the geological record. Fisher also took issue with Joly's reliance upon uniformity, with his inadequate allowance for the recirculation of sodium in the erosion-sedimentation cycle, and with his inadequate account of the origin of oceanic chlorine. But for all of his doubts, Fisher adjudged the method to be an important innovation in the search for the earth's age, and one worthy of serious consideration.[79]

None of the doubts seriously swayed Joly, however. He had gone into great detail, including comparative quantitative chemical analyses of the sodium found in primitive and sedimentary rocks, to show that recirculation must be a minimum, and he was convinced that airborne salt from the ocean winds was the only source of recirculated sodium worth considering. He had also discussed in quantitative detail the various chemical forms in which sodium and chlorine reached the sea, and was confident that an adequate balance had been accounted for. And he had made what he regarded as more than ample allowance for the assumptions involved in his calculations. Thus when he responded to his critics, he gave little ground indeed, and when he addressed the British Association in 1900, he defended his position with the same assurance and finality so often displayed by Kelvin himself.[80] In the end he was persuasive. Even if all of his critics were not satisfied, his argument became one of ths strongest weapons in the geologists' defense of the 100 million year time scale, and its influence lasted well into the new century.[81]

As the century drew to a close, there was an understandable urge among the historically conscious Victorians to sum up the labors of the era just past, and wherever possible to tie up loose ends and plot the path of the future. In the case of the age of the earth, however, the problem of tying up loose ends, that is of reconciling the conflicting demands of geology, physics, and biology, appeared as insoluble as ever. No one, of course, could predict the sudden change that the new century would bring, and the *fin de siècle* remarks of Geikie, Chamberlin, and Kelvin had, if anything, merely emphasized the apparent magnitude of the gap. Joly too, although his new hypothesis had ended the era on a note of promise for geology, had pointedly avoided any reference to his disagreement with the restrictions of the physicists. But for all the disagreement, the desire for unity was still strong, and thus it was not inappropriate that the task of summation before the British Association, the body that had provided the forum for so much of the controversy, fell to one of the strongest advocates of compromise.

William J. Sollas (1849-1936) had begun his professional career just as Kelvin's influence in geology began to take effect. He had therefore never accepted Lyellian uniformity as an operational postulate in geology—even in the modified form advocated by Geikie—and had been from the very first an advocate of both a short geological time scale and Huxley's geological evolutionism. As early as 1876 he had argued that, since the cooling of the earth and sun were inescapable facts, it was equally inescapable that geological processes, especially those dependent upon plutonic and meteorological action, must be slowing down. He was categorically unwilling, therefore, to accept any calculation of time which postulated uniform rates of geological change.[82] Although he gave his support on the question of time to the physicists, he never followed them slavishly; and while granting the inviolability of the physical laws upon which Kelvin's arguments were based, he remained free to question the physical evidence. Thus by 1900, he was quite ready to entertain a limit on the earth's age that was well below Geikie's 100 million years, or even Joly's 80 to 90 million, but he was unwilling to accept a result as low as the one Kelvin then demanded. When called upon to address the geological section of the British Association in the year following Geikie's dramatic remarks, therefore, he took on the role of Procrustes himself, paring and stretching the most prominent of the estimates of time until they were in apparent agreement. It was a fascinating exercise, because he strove consciously to evaluate each case with dispassionate balance, and yet his own convictions were always apparent.[83]

Sollas paid no attention to the more radical estimates of time on either side. The four estimates he sought to bring into harmony were George Darwin's 50 million years as the *minimum* age of the moon which Sollas, too, accepted as his minimum; Joly's 90 million years for the age of the oceans, which he took as a maximum; Geikie's 100 million years for the age of strata; and Kelvin's 24 million years. Of these, he made no attempt to alter Darwin's position. The laws upon which it was based he accepted unquestioningly, and the result was in perfect accord with those he obtained from the other arguments. Nor did he question the principles of Kelvin's secular cooling argument. Instead, he merely reviewed the diverse array of measurements of the earth's thermal gradient and concluded that Kelvin had made an unlikely choice on the basis of inadequate data. A better choice of the average gradient would raise Kelvin's result to 50 million years. He also endorsed Joly's method, but argued that two important factors had been overlooked. In the first place, by concentrating on a few major rivers, Joly had neglected the much higher alkaline content of many of the smaller rivers; and secondly, he had not made sufficient allowance for the much greater rate of geological activity in the past. These two factors together, Sollas believed, would reduce Joly's estimate by some 10 million to 30 million years. Turning finally to the question of sedimentation, he again took up the problem of uniformity, and while stressing the conviction that rates of change had varied, conceded that uniformity was a necessary working hypothesis in estimating the time of formation of strata. He therefore concentrated upon making those corrections necessary to account for the fact that the area receiving sediment was not equal to the area being denuded. Adopting an approach similar to Wallace's, he argued that sedimentary desposits accumulated near the mouths of rivers in areas far smaller than the river drainage basins. He than added several other corrections not mentioned by Wallace which would further hasten the deposition of the maximum thickness of sediment, and concluded that if Geikie were correct in assuming an average denudation of 1 foot in 2400 years, then the rate of maximum sedimentation would probably be 1 foot in 100 years. On this basis the age of all stratified deposits would be only 26 million years, a figure which he believed the biologists would come to accept once they rid themselves of their misconceptions about the time necessary for organic change, and one which provided a long Precambrian period when taken with the 50 million to 60 million years provided by the other estimates.

Sollas' address brought a fitting close to the nineteenth century's concern with the earth's age. Although certainly not entirely representative of geological thought in 1900, it consolidated many of the elements which had dominated the question for nearly four decades. His attack on "Lyellian" uniformitarianism while at the same time adopting uniformity as a necessary working principle, his acceptance of the principles of the physicists' arguments, his assurance that geology alone could provide the necessary methods for measuring time once it had adopted the practices of measurement and experiment, his belief that biology would ultimately take its time from geology, and finally his conviction that the earth's age must be limited to a period somewhere near Kelvin's original estimates, were all themes that had been repeated until they had become truisms. Not everyone was as confident as Sollas. Perry, Chamberlin, Darwin, and Geikie had all introduced doubts, and probably most geologists agreed with Joseph LeConte that the true solution to the problem of the earth's age would have to wait until the new century.[84] But no one really doubted that the solution would be forthcoming, probably soon.

Geology was confident. During the preceding decades, geologists had developed quantitative methods and amassed a wealth of data which they were sure would lead them to a true understanding of the history of the globe. It may indeed be justly argued that on the question of time thay had not progressed conceptually much beyond the positions of Lyell, Darwin, and Kelvin years earlier. And an even more critical assessment can be built around their failure to recognize many of the assumptions which had brought their results into such remarkable agreement. For the geologists themselves, however, it was significant that their results did agree, and that the earth's age was limited and definable. It was important, too, that these limits were no longer imposed by the calculations of the physicists but by the evidence of geology and paleontology. The geologists were confident that they had outlived the excesses of uniformitarianism and had made their science into a full and equal partner with the other exact sciences. Perhaps as important as all of this, many geologists were confident enough to question dominant hypotheses in both physics and geology.

The problem of geological time had played an important part in molding the geologists' attitude, and for this they willingly acknowledged a debt to Kelvin. They no longer accepted his limits on time, but they paid homage for the stimulus he had given their science. Of the multitude of tributes paid him at the end of the century, a few taken from geologists who had come to oppose his views seem accurately to reflect geological opinion. Geikie, for example, summed up much of this feeling in his contribution to Kelvin's obituary:

> Even those geologists, paleontologists, and biologists

> who most keenly dispute the validity of the arguments
> whereby he sought to compress the antiquity of the
> globe within limits that seem too narrow for the evolu-
> tion of geological history, must admit that in turning
> the brilliant light of his genius upon this subject he did
> a notable service to progress of modern geology.[85]

And Chamberlin in the heat of a point by point refutation of Kelvin's
arguments took time to remark:

> . . . however inevitable must have been the ultimate
> recognition of limitations [of time], it remains to be
> frankly and gratefully acknowledged that the con-
> tributions of Lord Kelvin, based on physical data, have
> been most powerful influences in hastening and guid-
> ing the reaction against the extravagant time postu-
> lates of some of the earlier geologists. With little doubt,
> these contributions have been the most potent agency
> of the last three decades in restraining the reckless
> drafts on the bank of time. Geology owes immeasura-
> ble obligation to this eminent physicist for the deep
> interest he has taken in its problems and for the pro-
> found impulse which his masterly criticisms have given
> to broader and sounder modes of inquiry.[86]

In crediting Kelvin with the overthrow of the excesses of midcentury
uniformitarianism, these latter day uniformitarians may have been less
than just to Lyell and his followers. His influence cannot be denied,
however. Even his original estimate of 100 million years was far from
dead. For although the chronological estimates which the geologists
defended were indeed drawn from geological evidence, they nonetheless depended upon assumptions which, consciously or unconsciously,
had been made in order to bring them within Kelvin's limits. And in
this respect his influence was far from over.

A Biological Postscript

Although the words of Geikie and Chamberlain aptly sum up the
attitude of geologists toward Kelvin and the age of the earth at the end
of the nineteenth century, the record would not be complete without
some notice being taken of the brief but enthusiastic final endorsement
of his views from the side of biology. For nearly three decades biologists
had taken little part in the ongoing debate over geological time. Evolu-

tion had become firmly established as a biological principle. Yet despite the fact that many geologists looked upon the fossil record as the most persuasive evidence for the earth's great antiquity, too little was known about the mechanism of organic change for it to be of any use as a measure of exact chronology. Most biologists, therefore, were content to follow Huxley's dictum and take their time from geology or even from physics. Indeed as late as 1898 Wallace was still willing to argue that time "even as limited by Lord Kelvin's calculations," would prove sufficient for the operation of natural selection.[87] Poulton spoke for a younger generation, and in welcoming Perry's attack upon Kelvin he no doubt reflected the discontent of many of his contemporaries. Few of them, however, followed his lead. Instead, one group of biologists, the saltationists, went well beyond Wallace in embracing even the most restrictive of Kelvin's results.

The saltationists' argument that organic change takes place through abrupt transformations, or mutations, rather than through minute variations had been entertained off and on for decades before it was given a sudden, renewed prominence by the rediscovery in 1900 of Gregor Mendel's neglected genetic theories. Hugo DeVries (1848-1935), the Dutch botanist who rediscovered Mendel's forgotten works, became the vocal leader of the new saltationist school. His aim was to replace the theory of natural selection by a theory of genetic mutation. And since mutations could account for species change within a relatively short time span whereas natural selection could not, he found the restrictions of Kelvin's chronology particularly congenial. Actually DeVries devoted only a few pages of his massive *Die Mutationstheorie* to the question of time, but what he said was quite significant since he stressed not merely that his theory was compatible with Kelvin's chronology, but that the physicist's restrictions provided conclusive *proof* that evolution *must* take place by mutation.[88] In his discussion DeVries consistently ignored any contemporary doubts about the earth's age and reviewed only the most restrictive estimates. Even when admitting that some vagueness was inevitable in such calculations, he calmly neglected any uncertainties that might enlarge the estimates cited. For example, Geikie and George Darwin were included in his list of authors whose numerical results were used to support a limited chronology, even though both men specifically denied that their calculations placed any maximum limit on the earth's age. By concentrating on numerical results and ignoring any reservations or underlying assumptions, DeVries was able to point to the apparent agreement among the calculations and assert emphatically that "20 to 40 million years is the period of the duration of life upon the earth."[89] Then in a marvelous piece of circular reasoning, he provisionally adopted Kelvin's 24 million years for the age of the earth; applied it to the hypothesis of mutation; assumed that a few thousand mutations would determine the elementary characters of a higher plant; concluded that a few thousand years must be the average time between periods of mutation; and finally asserted that a few million years would suffice for the whole evolution of the vegetable and animal kingdoms. As if it were the result of rigorous deduction he then proclaimed baldly that "the doctrine of mutation does not demand a longer period for the duration of life than that which has been given by LORD KELVIN, *viz.*, 24 million years."[90]

Despite the shortlived prominence of DeVries' biological theories, his position on the age of the earth was an anachronism. He received no support from geologists who were in revolt against Kelvin's restrictions. But more important, almost as soon as *Die Mutationstheorie* was published (1901) and long before the appearance of its English translation (1909), the newly discovered phenomenon of radioactivity had undermined the foundations of Kelvin's arguments.[91]

REFERENCES

[1]Salisbury (1894), *Presidential Address.*

[2]Volume II of Kelvin's *Popular Lectures and Addresses* appeared in February, and Huxley's *Discourses, Biological and Geological Essays* appeared in April 1894.

[3]This account is by Huxley's student, Henry Fairfield Osborne. (See: Huxley, L. (1900), *Life of Huxley*, II:376-377.)

[4]*Ibid.*, II:378.

[5]See, e.g.: Stewart (1896), *Conservation of Energy*, pp. 151-152; and Holm (1902), *New Solar Theory*, p. 166.

[6]Challis (1863b), *Source of the Sun's Heat*, p. 461.

[7]*Ibid.*, p. 465.

[8]Williams. W.M. (1882), *Current Discussions*, p. 211. The quotation is from a synopsis of his argument that Williams published in order to show that his ideas had been unjustly neglected.

[9]Williams, W.M. (1870), *Fuel of the Sun*.

[10]Marchant (1916), *Letters of Wallace*, I:263. Letter dated 14 May 1871.

[11]Siemens (1882), *Conservation of Solar Energy*.

[12]Croll (1868), *Geological Time*, 35:367-375. This argument is discussed in Chapter 3.

[13]Croll (1875b), *Climate and Time*, pp. 311-367.

[14]Croll (1877), *Origin and Age of the Sun*, p. 309.

[15]*Ibid.*, p. 317.

[16]*Ibid.*, p. 317-318. Croll's remarks were still being echoed sixteen years later in the words of C.D. Walcott: "Of the degree of the sun's heat we know so little that conjectures in relation to it have little force against the conditions indicated by the sedimentary rocks and their contained organic remains." Walcott (1893), *Geologic Time*, p. 356.

[17]Croll (1878c), *Origin of Nebulae*; Croll (1878a), *Age of the Sun*; Petrie (1878), *Age of the Earth*; Plummer (1878), *Age of the Sun*; Preston (1878), *Age of the Sun's Heat*.

[18]Croll (1886), *Climate and Cosmology*; Croll (1890b), *Stellar Evolution*.

[19]Kelvin (1889), *On the Sun's Heat,* pp. 410-412.

[20]Croll (1890b), *Stellar Evolution*, p. 68.

[21]Croll (1871c), *Age of the Earth*; Croll (1875b), *Climate and Time*, p. 335; Croll (1877), *Origin and Age of the Sun*, pp. 323-325; Croll (1876a), *Tidal Retardation Argument*.

[22]Darwin, G. (1886), *Presidential Address*, pp. 512-518.

[23]*Ibid.*, p. 518.

[24]Fisher (1881), *Physics of the Earth's Crust*.

[25]A. R. Wallace, for example, was convinced that Fisher had proven Kelvin wrong on the question of the earth's rigidity and praised *The Physics of the Earth's Crust* as "really a grand book . . . though full of unintelligible mathematics." Marchant (1916), *Letters of Wallace*, II:74.

[26]Fisher (1882), *Depression of Ice Loaded Lands*.

[27]Fisher (1893), *Rigidity not to be Relied upon*; Becker (1893), *Fisher's New Hypothesis*; Barus (1893b), *Criticism of Mr. Fisher's Remarks*.

[28]Reade (1878a), *Geological Time*, pp. 234-235.

[29]Reade (1878b), *Age of the World*, p. 146.

[30]*Ibid.*, p. 150.

[31]Reade (1883a) *Examination and Calculation*; Wallace (1883), *Reply to T. Millard Reade*; Reade (1883b), *Reply to Mr. Wallace*.

[32]Geikie, A. (1892), *Presidential Address*, p. 192.

[33]*Ibid.*, pp. 185-186.

[34]Woodward, R.S. (1889), *Mathematical Theories*, pp. 61-64. Quotation, p. 63.

[35]Walcott (1893), *Geologic Time*, pp. 343-344.

[36]Darwin, G. (1886), *Presidential Address*, p. 571; Williams, H. Shaler (1893a), *Geological Time Scale*, pp. 292-294; Prestwich (1895), *Collected Papers*, pp. 47-48.

[37]Bonney (1893), *Story of Our Planet,* pp. 481-485.

[38]McGee (1892), *Comparative Chronology*.

[39]An abridged report of McGee's remarks appear in McGee (1893), *Note on the Age of the Earth*.

[40]Edward B. Poulton reported this opinion of Perry's as a result of a conversation during the British Association meeting of 1890. At that time Perry refused to challenge Kelvin. Poulton (1896), *A Naturalist's Contribution*, pp. 9-10.

[41]Perry (1895a), *Age of the Earth*, p. 224.

[42]The letter was evidently widely circulated before publication. Perry mentions sending it to Joseph Lamor, Oliver Lodge, Osborne Reynolds, Oliver Heaviside, and G.F. Fitzgerald. He also must have sent copies to Kelvin and Tait, and I have found copies in other archival collections including those of T.H. Huxley and H.A. Rowland.

[43]Perry (1895a) *Age of the Earth*, pp. 224-225. Letter dated 14 Oct. 1894.

[44]*Ibid.*, p. 226. Letters dated 22 Nov. and 26 Nov. 1894.

[45]*Ibid.*, p. 227. Letter dated 29 Nov. 1894.

[46]*Loc. Cit.* Letter dated 13 Dec. 1894.

[47]Kelvin (1895), *Age of the Earth*; Thompson (1910), *Kelvin*, II:941-943.

[48]Perry (1895b), *Age of the Earth*, pp. 582-585.

[49]Geikie, A. (1895), *Twenty-Five Years*, p. 371.

[50]Sollas (1895), *Age of the Earth*, p. 533.

[51]Fisher (1895), *Age of the World*.

[52]Haeckel (1899), *History of Creation*, I:131.

[53]Huxley (1909), *Discourses, Biological and Geological*, p. ix. From the preface dated April 1894.

[54]Poulton (1896), *A Naturalist's Contribution*.

[55]*Ibid.*, p. 43.

[56]Kelvin (1899), *Age of the Earth as an Abode Fitted for Life*.

[57]*Ibid.*, pp. 219, 225.

[58]Chamberlin, T.C. (1899), *Lord Kelvin's Address*, 9:891.

[59]Chamberlin, T.C. (1965), *Multiple Working Hypotheses*, p. 755.

[60]The first presentation of the *Planetesimal* hypothesis was at the British Association meeting in Toronto on 20 August 1897, only two months after Kelvin's address to the Victoria Institute. At that time it contained only a brief mention of the implications for Kelvin's chronology. (See: Chamberlin, T.C. (1897), *A Group of Hypotheses*.) For an interesting insight into Chamberlin's ideas on multiple hypotheses and the nebular theory see: Winnik (1970), *Science and Morality*.

[61]Chamberlin, T.C. (1899), *Lord Kelvin's Address*.

[62]*Ibid.*, 10:12.

[63]Geikie, A. (1895), *Twenty-Five Years*, p. 369.

[64]Geikie, A. (1899), *Presidential Address*.

[65]Geikie's use of the term "uniformity" came to be much closer to the "actualism" of the continental geologists, rather than the "uniformitaranism" of Lyell's followers. For a careful discussion of the distinctions see: Hooykaas (1963), *Principle of Uniformity*.

[66]Geikie, A. (1899), *Presidential Address*, p. 727.

[67]Geikie's brief history including its oversimplifications seems to have provided the basis for most subsequent summaries of the controversy over the age of the earth.

[68]Geikie, A. (1899), *Presidential Address*, p. 726.

[69]*Ibid.*, p. 727.

[70]Goodchild (1896), *Some Geological Evidence*.

[71]*Ibid.*, p. 259.

[72]*Ibid.*, pp. 271-272. Two years later Goodchild examined the physicists' arguments in still greater detail with the result that he became even more convinced of the questionable nature of their assumptions. (See: Goodchild (1897-1901), *The Earth's Internal Heat*.)

[73]*Ibid.*, p. 306.

[74]Joly (1899), *Estimate of the Geological Age*. The idea of determining the age of the oceans from their salt content actually originated with Edmund Halley in the eighteenth century, but was never tried until Joly independently developed a slightly different approach at the end of the nineteenth. (See: Halley (1714-16), *The Cause of the Saltiness of the Ocean*; and Joly (1915), *Birth-Time of the World*, p. 13.)

[75]Joly (1899), *Estimate of the Geological Age*, p. 247.

[76]*Ibid.*, p. 282.

[77]Sollas (1900), *Evolutional Geology*, pp. 301-303.

[78]Mackie (1898-1905), *The Saltiness of the Sea*.

[79]Fisher (1900), *John Joly's Estimate*.

[80]Joly (1900c), *On the Geological Age of the Earth*. Other examples of Joly's response to criticism include: Joly (1900b), *The Geological Age of the Earth*; and Joly (1901), *The Circulation of Salt and Geological Time*.

[81]An interesting indication of the spread of Joly's method is found in a similar calculation made on the continent by Eugene Dubois and based upon the circulation of calcium carbonate. The use of $CaCO_3$ was in some ways more closely related to the calculations of Mellard Reade in the seventies; but in presenting his initial postulates and devloping his argument, Dubois' approach was closer to Joly than to Reade, both of whom he consulted for data but whom he did not mention with regard to method. His result, moreover, was quite different from Reade's. He required only 45 million years as the *minimum* time for the total formation of strata. Dubois was convinced, however, that actually a much longer time must be allowed, and although he had no real data for calculation he did not consider even a *billion* years excessive. In this Dubois provides a classic example of going beyond limited data because of his conviction that more time was needed. (See: Dubois (1901), *Circulation of Carbonate of Lime*.)

[82]Sollas (1877), *Evolution in Geology*.

[83]Sollas (1900), *Evolutional Geology*.

[84]LeConte (1900), *A Century of Geology*, p. 278.

[85]Included in Lamor (1908), *William Thomson*, p. lxxxi.

[86]Chamberlin, T.C. (1899), *Lord Kelvin's Address, 9*:890. Similar opinions are to be found in: Darwin, G. (1886), *Presidential Address*, p. 518; Woodward (1889), *Mathematical Theories*, p. 61; and Harker (1915), *Geology in Reaction*, p. 107.

[87]Marchant (1916), *Letters of Wallace*, II:74-75.

[88]DeVries (1909), *Mutation Theory*, II:663-674.

[89]*Ibid.*, II:669.

[90]*Ibid.*, II:674.

[91]The discovery of radioactive heating early in 1903 cast severe doubts upon several of Kelvin's fundamental assumptions. (See chap. 6.) Nonetheless, DeVries still continued to place great emphasis upon the limitation of time for several years afterward, disregarding all evidence to the contrary. (See: DeVries (1904), *Evidence of Evolution*, and DeVries (1906), *Species and Varieties*, pp. 710-714.)

VI
Radioactivity and the Age of the Earth

March 16, 1903, marked the beginning of a new and radically different era in the development of theories about the age of the earth. On that day Pierre Curie (1859-1906) and his young assistant, Albert Laborde, announced the discovery that radium salts constantly release heat.[1] They had discovered the "unknown source of energy" that Kelvin had been unable to foresee in 1862. Perhaps the new age properly began with Henri Becquerel's (1852-1908) discovery of radioactivity in the spring of 1896,[2] but the geological significance of the new discovery was not recognized until Curie and Laborde made their startling announcement. The reaction was almost instantaneous. Once the discovery was announced, physicists and geologists quickly seized upon it as a possible, even probable, solution to the problem of the heretofore unreconcilable divergence between the physical and geological estimates of the earth's age.

The frantic activity which radioactivity stimulated in experimental physics was reflected in the speed with which the new discoveries were applied to the question of time, and the leader in the new quest soon emerged. He was Ernest Rutherford (1871-1937), the young professor of physics at McGill University in Montreal. Even before the announcement by Curie and Laborde, Rutherford and Frederick Soddy (1877-1956), an Oxford educated chemist also at McGill, had discovered the enormous energy associated with the release of alpha radiation from any known radioactive material.[3] Immediately, Rutherford and another associate, Howard T. Barnes (1873-1950), his future successor at McGill, began to examine the nature, magnitude, and source of the radioactive heating effect. By October, 1903, they were ready to announce that the production of heat directly accompanied and was quantitatively proportional to the number of alpha rays emitted by a substance.[4] And by early the following year, Rutherford had become convinced that the atoms of all radioactive elements, and perhaps all elemental atoms, contained enormous stores of latent energy.[5] From there it was but a short step to the conclusion that this new discovery must vitiate Kelvin's limitations on the age of the earth.

The opportunity to announce his conclusions proved dramatic, at least for Rutherford himself. Visiting England in the spring of 1904 to discuss his discoveries, he was invited to address a full audience at the Royal Institution. Some time later he provided this account of the lecture:

> I came into the room, which was half dark, and pres-
> ently spotted Lord Kelvin in the audience and realized
> that I was in for trouble at the last part of the speech
> dealing with the age of the earth, where my views
> conflicted with his. To my relief, Kelvin fell fast asleep,
> but as I came to the important point, I saw the old bird
> sit up, open an eye and cock a baleful glance at me!
> Then a sudden inspiration came, and I said Lord
> Kelvin had limited the age of the earth, *provided no new*
> *source of heat was discovered*. That prophetic utterance
> refers to what we are now considering tonight,
> radium! Behold! the old boy beamed upon me.[6]

The part of Rutherford's speech that had made him anxious came when he asserted that the earth could no longer be treated as a mere cooling body. Since it had been shown that radioactivity was constantly replenishing terrestrial heat by an unknown amount, there was no way of determining the rate of change, if any, in the earth's thermal gradient. It may, in fact, have been sensibly constant, and the terrestrial surface capable of supporting life for an extremely long time. Kelvin's old calculation could therefore provide no more than a *minimum* estimate of geological time at best. Furthermore, since radium might also be found to replenish solar energy, his estimate of the sun's age might require a similar revision. Thus Rutherford concluded:

> The discovery of the radio-active elements, which in
> their disintegration liberate enormous amounts of
> energy, thus increases the possible limit of the dura-
> tion of life on this planet, and allows the time claimed
> by the geologist and biologist for the process of
> evolution.[7]

Kelvin was far from ready to concede all of Rutherford's points, however. Radium fascinated him, and he discussed it at length with Rutherford and with the Curies, who visited England just after the heating effect had been announced.[8] But Einstein's announcement of special relativity was still a year away (it probably would have made as little an impression on Kelvin as it later did on Rutherford in any case) and Kelvin was still firmly tied to mechanical explanations and rigid conservation laws. Shortly after Rutherford's lecture, therefore, he published a note rejecting the idea that radium could perpetually emit heat, arguing instead that the energy emitted must be supplied to the radium by an external source. "I venture to suggest that somehow ethereal waves may supply the energy to the radium while it is giving out heat to the ponderable matter around it."[9] This was in effect the same suggestion that Curie had tentatively made in announcing his discovery, but which he never followed up.

According to Rutherford's biographer, A.S. Eve, Kelvin publicly abandoned this position during the British Association meeting later the same year, and this report seems supported by J.J. Thomson's account of a conversation in which Kelvin conceded that the discovery of radium had made some of his assumptions untenable.[10] But no retraction was ever published. On the contrary, two years later in 1906, the *Times* carried a series of letters in which Kelvin took on several of his more eminent younger colleagues — including Soddy, Oliver Lodge, A.S. Eve, and R.J. Strutt — in a public debate over the nature of radioactivity. Among other things, he once again specifically denied the contention that radium could account for the heat of the earth and sun, and left no doubt that in his opinion, gravitation remained the only possible source of such vast stores of energy.[11] He never again wavered from this position. Turning to it finally in one of the last papers before his death, he attempted again to explain the heat emission of radium by purely mechanical means. Ignoring the results of Rutherford and Barnes, he asserted that the heat produced by radium must accompany the expulsion of electrons rather than alpha particles, and treated the electrons as if they were propelled by some kind of spring mechanism within the atoms. The atoms would be "loaded" by absorbing energy by conduction and radiation from the surrounding ether, and thus the "process can go on forever, without violating the law of conservation of energy, and without any *monstrous* or *infinite* store of potential energy in the loaded Radium."[12]

If Kelvin died unregenerate, he was not alone. As Max Planck remarked: "A new scientific truth does not triumph by convincing its opponents and making them see the light, but rather because its opponents eventually die, and a new generation grows up that is familiar with it."[13] So it was with the question of the earth's age. After the initial enthusiasm for the possibilities raised by radioactivity had passed, several of the more notable spokesmen for the older generation returned to the conventional ways of estimating time. A few continued to fight for decades against the new results introduced by physics. But the vitality of the new science never slackened, and bit by bit the old conceptions of time and the old methods of calculating the earth's age gave way to an entirely new approach.

Kelvin Overthrown: Radioactive Heating

Even before Rutherford and Barnes completed their findings on the nature of radioactive heating, speculations had already begun to appear concerning the implications of the new discovery for solar physics and the age of the earth. In July, 1903, barely four months after the announcement by Curie and Laborde, *Nature* carried the first such speculation. It came from William E. Wilson (1851-1908), an independent gentleman-astronomer with a long standing interest in solar heat. Starting from the suggestion that the heat generated by radium might somehow "possibly afford a clue to the source of the energy in the sun and the stars," Wilson used a simple preliminary calculation to show that a density of radium as small as 3.6 grams per cubic meter would supply all of the heat currently radiated by the sun.[14]

Wilson's note apparently went unnoticed at first, but a few months later, *Nature* carried a second letter containing a similar speculation. This time the author was George Darwin, and the suggestion attracted immediate attention. Darwin had been publicly questioning the precision of Kelvin's chronology since 1886 and privately debating his assumptions for longer still; but as late as 1897, he still saw their differences as matters of degree rather than substance.[15] He now saw in the new discovery the vindication of his growing doubts, and quickly rose "to point out that we have recently learnt of the existence of another source of energy, and that the amount of energy available is so great as to render it impossible to say how long the sun's heat has already existed, or how long it will last in the future." But though he declared that Kelvin's calculations could no longer be considered valid, his own expansion of the time scale still remained quite conservative. "I see no reason," he said, "for doubting the possibility of augmenting the estimate of solar heat as derived from the theory of gravitation by some such factor as ten or twenty."[16] Within a week Darwin's suggestion was seconded by Joly, who had been pondering the question since Wilson's brief note first appeared. If anything, Joly was even more emphatic than Darwin. Although conceding that many questions still remained to be resolved by experiment, he left no doubt that "The gross dynamical supply of solar heat must no longer be regarded as affording a major limit both to solar age and geological time."[17]

For the next several weeks almost every issue of *Nature* contained letters speculating on the relationship between radioactivity and solar heat, and only one raised any serious doubts about the reality of the connection. William B. Hardy (1864-1934), a Cambridge biologist and colloid chemist, raised the question of the absence of Becquerel radiation (beta and gamma rays) in the solar spectrum. If radium were truly the source of the sun's heat, he argued, such radiation should be present in the solar spectrum, and yet all of his attempts to detect it had proven completely negative.[18] Immediately both Joly and Robert J. Strutt responded. Strutt (1875-1947), the son of Kelvin's old friend Lord Rayleigh, showed by a simple calculation that any beta and gamma radiation arriving from the sun would be filtered out by the atmosphere long before it could reach the earth's surface. "Even if all the sun's heat were due to radium," he concluded, "there does not appear to be the smallest possibility that the Becquerel radiation from it could ever be detected at the earth's surface."[19] Joly echoed Strutt's conclusion and added several other reasons for assuming the existence of radium on the sun. Among others he noted the presence of radium on the earth, the presence of helium (which was coming to be ever more closely associated with radioactivity) on the sun, and perhaps most important, the fact that the duration of solar heat calculated from purely dynamical sources ran "counter to geological data."[20] By the end of 1903 it seemed clear to several investigators that radioactivity must in some way be responsible for the heat of the sun. This belief was based entirely upon speculation, however, and even in the enthusiasm generated by the new discovery, no one was unaware of the problems remaining before it could be verified.

The speculations raised in the notes to *Nature* were pursued still further during the next several years, but nothing new was added to the original hypothesis. Many basic questions concerning radioactivity and the sun continued unanswered and unanswerable, and as late as 1915 Kelvin's solar theory still had its defenders.[21] But despite the lingering uncertainties, by 1905 many, if not most, physicists and geologists had accepted the idea that radioactivity would ultimately prove to be responsible for the sun's heat. After all, as both Rutherford and Strutt pointed out, the presence of helium in the solar spectrum did give convincing evidence that radioactivity was present in the sun. Thus the simple gravitational explanation of solar energy could no longer be considered adequate. Indeed, Rutherford went so far as to speculate that the sun's enormous heat might itself somehow amplify the rate of radioactive change and thus greatly extend the probable limit of its age.[22] Other commentators were quite willing to concur.[23]

The problem was far from settled, however. As the study of radium and other radioactive materials progressed, it became increasingly apparent that radium could not directly account for the sun's energy, and yet no alternative seemed possible. Thus, by the end of the decade, the importance a scientist was willing to give the radioactive theory of the sun's heat depended in great measure upon the value of the earth's age that he accepted from other arguments. For example, Joly, who advocated a comparatively short terrestrial antiquity on the basis of his sea salt theory, argued that "the radio-thermal theory cannot be supposed to account for any great part of solar heat unless we are prepared to believe that a very large percentage [60%] of uranium can be present in the sun, and yet yield but feeble spectroscopic evidence of its existence." Such an assumption, he felt was unnecessary to allow ample time for the duration of the sun and earth.[24] Rutherford, on the other hand, saw the possibility of radioactivity enormously increasing the allowable age of the sun. He held this view, moreover, even though he had to admit that a simple calculation showed that even if the sun were entirely composed of uranium in equilibrium with its products, it could generate only one-fourth of its lost heat. He continued, however, by saying, "At the enormous temperature of the sun, it appears possible that a process of transformation may take place in ordinary elements analogous to that observed in the well-known radio-elements." In which case, he concluded, "The time during which the sun may continue to emit heat at the present rate may be much longer than the value computed from ordinary dynamical data."[25] The difficulties presented by the problem of accounting for the sun's enormous expenditure of heat remained unsolved until the discovery of nuclear fusion over two decades later. Meanwhile, although occasional references to the problem were made, it faded into the background as far as geological time was concerned, while attention was focused on the more accessible question of radioactivity and the earth's thermal loss.

Joly was the first to turn to the direct terrestrial implications of radioactive heating. In a letter to *Nature* in October, 1903, he raised the possibility that radioactivity might be partially responsible for the earth's thermal gradient as well as the sun's heat. His intention was clear. If radioactivity could be shown to contribute to the earth's secular heat loss, Kelvin's calculation of time would have to be extended, and thus "the hundred million years which the doctrine of uniformity requires may in fact yet be gladly accepted by the physicists."[26] Although Joly's letter appeared only a month after George Darwin's note on the possible relationship between radioactivity and solar heat, it failed to draw the same immediate response. Then, early in 1904, Julius P. L. J. Elster (1854-1920) and Hans F. K. Geitel (1855-1923), long term collaborators at the gymnasium at Wolfenbuttel, announced the startling discovery that common garden earth contains a surprisingly high density of radium.[27] Within months both Rutherford and C. Liebenow, had calculated that radium alone, in the density found by Elster and Geitel, would more than account for the earth's entire heat loss through surface conduction.[28] Indeed, Rutherford's calculation raised the possibility that the earth might not be cooling at all, but rather that it might actually be self-heating. He was not willing to push his conclusions that far at the time, however, and instead limited himself to asserting that the present terrestrial surface thermal gradient may have remained constant for an indefinite period of time. Nonetheless, that conclusion in itself was sufficient to cast doubt upon Kelvin's or any other calculation of time based upon the earth's secular cooling.

Frederick Soddy, Rutherford's former collaborator, was not so restrained in his judgment. After his return to England in 1903, Soddy had entered into an informal competition with Rutherford over who would first publish a book on radioactivity. Both works appeared in 1904, and while Rutherford's was the more substantial, Soddy's took the more dramatic stand on the question of the earth's age. "The really essential advance that the results dealt with in this book lead to," he asserted near the conclusion, "is that the limitations with respect to the past and future history of the universe have been enormously extended."[29] He particularly took Tait and Kelvin to task for restraining the "legitimate aspirations of science" by their dogmatism. Taking up each of Kelvin's arguments in turn, Soddy declared that radioactivity vitiated even the most exact of his results. The possibility that the earth might be self-heating rather than self-cooling meant that its age was limited only by the maximum time necessary for radioactive decay to produce the present concentration of radioactive ores, not by the time for mechanical cooling. The geologists' 100 million years might prove long enough, he admitted, but his own calculations pointed to a result closer to one billion years. Furthermore, even this result might be increased if it could be shown that somehow heavier elements were generated from lighter ones. Soddy knew that he was upon very speculative ground here, for although the process that he envisaged might allow for successive cycles of atomic regeneration and decay and thus indefinitely extend the limits of the earth's age, it could do so only at the expense of rejecting the second law of thermodynamics in the subatomic realm. For a time Soddy was willing to entertain such a possibility, and citing Maxwell's analogy of the "sorting demon" in its behalf, asserted that it provided a more complete and satisfactory view of the state of the universe than any as yet proposed.[30] Nonetheless, except for R. K. Duncan, he received little support in this the most speculative of his suggestions,[31] and he pursued it no further.

As Rutherford and Soddy moved into other inquiries, the task of investigating the earth's surface radioactivity fell to Strutt, who tried to extend and refine what he considered to be the inadequate findings of Elster and Geitel. Strutt quickly discovered, however, that as he improved the accuracy of his data, the anomaly in attributing the earth's heat to radium also increased. Even disregarding the effects of uranium and thorium, his analyses of igneous rocks revealed 50 to 60 times more radioactivity than the average concentration needed to maintain the earth's temperature. The anomaly seemed explicable only if radioactivity were a surface phenomenon confined to exposed rocks or if all radioactive ores were for some reason concentrated in the crust. In 1905 Strutt leaned to the former conclusion, but more careful study the following year led him to the latter. Accepting this conclusion, his calculations indicated that all of the earth's radioactivity must be confined to a thin crust about forty-five miles deep.[32] This unexpected result, which Strutt himself considered to be incredible but inescapable, raised numerous questions for both geology and physics which could not be answered immediately. But wherever these questions might lead there seemed to be little doubt that Kelvin's purely mechanical explanation of terrestrial heat would never again be tenable.

The geological reaction to the new discoveries was decidedly mixed. To some, radioactivity seemed a threat. Sollas, for example, reluctantly conceded that Kelvin's view of the earth's age was no longer tenable, but nonetheless looked upon radium as an apparition, "vague and gigantic, threatening to destroy all faith in hitherto ascertained results, and to shatter the fabric of reasoning raised upon them."[33] Others welcomed each discovery joyfully. T. Mellard Reade, despite his belated acceptance of Kelvin's conclusions only a decade earlier, saw the shackles being stripped from geology. "The bugbear of a narrow physical limit of geological time being got rid of," he exclaimed, "we are now free to move in our own field of science."[34]

Between these extremes lay all shades of opinion. Geikie no doubt represented many of the older generation when he first welcomed radioactivity as a way out of the impasse imposed by physics, subsequently came to doubt its value, and then returned to the old theory of gradual thermal loss.[35] George Darwin, on the other hand, was more typical of younger men who looked for a degree of flexibility in the older theories, and in his own case readily declared that his theory of lunar genesis would probably accommodate an earth 500 million to 1,000 million years old.[36] There were occasions, too, such as the extended debate over terrestrial temperatures which was carried on in the United States in 1908, when the whole question of radioactivity was ignored.[37] But perhaps the most interesting discussions came from a small, influential band, led most vocally by Joly, who continued to see radioactivity as a means of reconciling Kelvin's later results with the 100 million years that had come to be accepted by geology, but who refused to concede the longer time spans that were soon to be demanded by the results of radioactive dating.[38] Joly and his supporters saw radioactivity as an extension rather than a refutation of Kelvin's arguments, but more than that, they exemplified the growing assurance among geologists that their own methods might well provide a more accurate measure of the earth's age than the much vaunted mathematics of the physicists. Thus for several decades they kept the nineteenth century conception of time alive while providing a valuable check upon the speculations of the new generation of physicists.

Radioactive Dating

As discovery followed discovery, the possibilities opened up by radioactivity seemed almost limitless, and the predominantly young investigators eagerly challenged even the most venerable established conclusions. Indeed, by the time Joly issued his word of caution in 1908, the physicists had already announced discoveries which went well beyond the thermal properties of radium and opened the way to speculations about an ever expanding time scale. The discoveries and accompanying speculations came with astonishing speed. Rutherford and Soddy first pointed the way in late 1901 when they turned from the investigation of radioactive radiations—which had dominated research since Becquerel's original discovery — to the examination of the radioactive materials themselves and the products of radioactive decay. By 1904, Rutherford, by then working independently of Soddy, had become convinced that helium was somehow produced by radioactivity. This assumption was yet to be verified experimentally; but in keeping with the spirit of the times, he immediately suggested that the helium trapped in radioactive minerals might provide a means for determining the age of the earth.[39] Three years later, his friend and sometimes collaborator, Bertram B. Boltwood (1870-1927) of Yale, made a similar suggestion with regard to the lead found in radioactive ores.[40] Neither man pursued his own suggestion, but others did, and both methods produced results far in excess of the ages accepted by most geologists.

In retrospect, the concept of radioactive dating seems simple. One merely determines the end product of a radioactive decay series and the time necessary to produce a unit amount of that product from the decay of the parent element; then for a given mineral, the ratio of the amount of the end product it contains to the amount of the original parent element, multiplied by the time necessary to create a unit amount of the product will give the age of the sample. In practice, of course, this procedure is not nearly so simple, as the numerous methods designed to refine it clearly indicate. And for the scientists who first developed the method, the problems, unknowns, and uncertainties were formidable indeed. In the first place, neither the end products nor the rates of radioactive decay were known with any degree of certainty; and secondly, the enormous times involved made it extremely difficult to determine either of these factors experimentally with any degree of precision. For a while, in fact, there was considerable debate over whether stable elemental end products were actually formed at all.

The first potential product of radioactive decay to be seriously investigated was radium, which may be used as an example of the problems encountered in the early years of investigation. The question of the origin of radium arose first in 1902 as a result of the investigations of Rutherford and Soddy on radioactive transformations. After over a year of research they had identified several short-lived intermediate products which they were unwilling to call elements (the transformation of *elements*, after all, smacked of alchemy), and for which they suggested the term *metabolon*. They could not get away from the fact, however, that radium itself, although universally regarded as an element, decays far too rapidly to be as old as the minerals in which it is found. The only solution seemed to be that radium too must be a *metabolon* with a "life" of a few thousand years.[41]

The conclusion that radium must continually be created, probably as a disintegration product of uranium or thorium, proved to be difficult to substantiate, however. In early 1904, Soddy, by then working with William Ramsay in England, reported negative results in his experiment to determine if radium is a product of uranium.[42] This report, in turn, prompted Joly to suggest that radium might in fact be the result of a combination rather than a decomposition of elements. But massive experimental difficulties barred any further investigation along this line.[43] In the meantime, both Boltwood and Strutt had independently measured the relative abundance of uranium and radium in several minerals and found their ratios to be constant.[44] Nonetheless, although this was strong evidence that some relationship existed between the two elements, it provided nothing definite to show that uranium was the "parent" of radium. Then in 1905, Soddy repeated his experiment under better conditions and reversed his earlier conclusions.[45] He was now convinced that radium was a product of uranium decay. Conviction was not proof, however, and subsequent experiments run independently by Boltwood and W. C. D. Whetham gave negative and positive results respectively.[46] The key to the difficulty lay obviously in the still unknown factors affecting the experiments, and despite his negative results, Boltwood refused to dismiss the possibility that radium was the product of uranium. Instead he argued that other products with slow decay times must precede and thus delay the formation of radium. Such a conclusion would mean that Soddy's experiment could not be correct, but it seemed to Boltwood the only possible way of explaining the constant proportionality between uranium and radium. It was actually not until 1909, after six years of experimentation, that Soddy was able to show, more or less certainly, that radium is a product of uranium decay.[47] But even without the rigor of this proof, the evidence for a connection between uranium and radium was sufficiently strong for it to be generally accepted by most investigators in the field.[48]

Both helium and lead, the first two elements to be used for radioactive dating, raised questions about their roles in radioactive decay similar to those raised by radium. In 1903 Soddy and William Ramsay (1852-1916) announced the discovery of unmistakable signs of helium in the spectra of the decay products of radium bromide;[49] and by early the following year Rutherford had become convinced that helium was invariably associated with radioactivity. It took another four years, however, for the nature of the connection to be established. During 1905 and 1906, both Rutherford and Strutt independently found evidence that the helium produced by a radioactive transformation was associated with the emitted alpha particles rather than with the end products of decay, but they were not yet ready to assert that the alpha particle *was* itself a helium ion.[50] A serious problem arose, moreover, because the alpha particles from uranium, radium, actinium and thorium all appeared to be identical, whereas Boltwood was convinced that helium was in no way connected with thorium decay.[51] Thorium was in fact to be the source of numerous difficulties over the next several years. It had already led Boltwood to dismiss the discovery of radiothorium by William Ramsay and Otto Hahn as "a new compound of Thorium-X and stupidity,"[52] and in this case it led to a serious disagreement between him and Strutt.[53] But the fact remained that if Boltwood proved right in this instance, the role of helium in radioactive decay would remain in doubt. The uncertainty lasted only until 1908 when the direct connection between helium and alpha particles was finally established by Rutherford, Hans Geiger (1882-1945), and Thomas D. Royds (1884-1955) at Rutherford's new laboratory in Manchester.[54] By that time, however, Strutt had already begun to develop techniques for helium dating, and so the new discoveries produced nothing new other than to give an added sense of validity to his conclusions.

Lead in many ways presented an even more perplexing problem than helium. In the first place neither the existence of isotopes nor the concept of atomic number, both of which were critical for the identification of the products of radioactive decay, were known during the first decade of the century. And secondly, thorium continued to perplex investigators, and to raise serious doubts about whether lead was a product of thorium decay. Boltwood in particular was distrustful of the results obtained from thorium, although he devoted considerable effort to unraveling its riddle. His frustration is apparent even in these lighthearted comments sent to Rutherford in December 1905:

> I am beginning to believe that thorium may be the
> mother of that most abominable family of rare-earth

elements, and if I can lay the crime at her door I shall make efforts to have her apprehended as an immoral person guilty of lascivious carriage. In point of respectability your radium family will be a Sunday school compared with the thorium children, whose (chemical) behavior is simply outrageous. It is absolutely demoralizing to have anything to do with them.[55]

Boltwood was never able to resolve his doubts, nor were his immediate successors. Even after more than two decades of labor the inconsistencies in the thorium/lead ratio remained too great to be depended upon, although there seemed no alternative to the conclusion that lead was a product of thorium decay.[56]

In spite of the difficulties, however, Boltwood's discovery that lead is invariably associated with all uranium ores led him to suggest in 1905 that it might be the end product of the uranium but not the thorium transformation series.[57] In the same year Soddy observed that an alpha particle emitted from polonium would reduce the atomic weight of the atom from 212 to 208, a value very near that of bismuth (208.5) and lead (206.9); but, unlike Boltwood, he was not yet ready to speculate on whether either of them was the ultimate product of the decay series.[58] Rutherford thereupon pointed out that the emission of eight alpha particles — if they proved indeed to be helium atoms — would reduce the atomic weight of uranium from 238.5 to 206.5, a value even closer than Soddy's to the atomic weight of lead. That result, plus Boltwood's analyses of the presence of lead in uranium ores, convinced him of at least the probability that lead was the stable end product of uranium decay.[59] The evidence for this conclusion was entirely circumstantial, however, and so it remained for another decade. Even after the experimental demonstration of the production of lead by uranium, the inconsistencies in the thorium/lead ratios continued to plague researchers. Yet with all of its drawbacks, the evidence was sufficiently strong by 1906 for physicists and geologists to feel free to proceed on the assumption that lead would prove to be the product of uranium and to qualify their results accordingly.

Thus, in spite of the admitted uncertainties surrounding the roles played by helium and lead in radioactive decay, the scientists who recognized their implications for the age of the earth — in all a rather small group composed primarily of physicists but including a few geologists and chemists — did not hesitate to apply the knowledge that was available to the problem of geochronology. For the most part, these investigators were convinced that radioactive heating had overthrown Kelvin's arguments, and they looked to radioactivity to supply the new key to a quantitative measure of the earth's age. Rutherford, it seems, provided the germ of the idea in 1904 when he suggested the possibility of using helium trapped in radioactive minerals to determine their age. By 1906 he had measured the helium content of a few selected ores and had estimated their ages at a minimum of 400 million years.[60] Even so he had already been anticipated by Strutt, who a year earlier had measured the helium content of a radium bromide salt and had calculated its age at two *billion* years.[61] Needless to say, these results greatly exceeded the most liberal of Kelvin's estimates and those of the majority of geologists as well.

Rutherford continued to speculate and to make seminal suggestions concerning radioactive dating, but his major researches led him in other directions. Consequently it was Strutt, newly appointed to the chair of physics at Imperial College, who was primarily responsible for developing the techniques of helium dating. Working, as was his custom, almost entirely alone, Strutt spent several years investigating the appearance of helium in various minerals and attempting to calculate their ages. He knew, of course, that his methods could at best give only minimum values for the ages of his samples, since some of the helium generated by radioactive decay must have escaped with time. He was hampered, too, by a lack of exact data which forced him to rely upon estimates (or educated guesses) for the number of helium producing transformations in the uranium-radium decay series. Nonetheless, between 1908 and 1910 he tested the helium content in a number of minerals, calculated their ages, and related his results to their stratigraphic locations. His aim was twofold: to find out if helium actually accumulates in minerals with time, and if so, to use it to determine their ages.

The problems were numerous. At the outset Strutt found that secondary and tertiary minerals were generally not amenable to radioactive measurements, and so he turned for his initial experiments to phosphate nodules and fossil bones, both of which were rich in radium and had clearly defined geological ages. Even so, his initial results gave a disappointingly poor correlation between helium content and relative geological ages. The discrepancy was quickly explained, however, and a new problem introduced when a subsequent experiment showed that under laboratory conditions the sample nodules lost helium at a faster rate than radium generated it. Clearly, any helium trapped in the samples was fortuitous.[62] The next step, therefore, was to analyze progressively harder minerals which might retain helium more efficiently; while at the same time searching for a direct method of determining the rate of helium production in ores. Although anticipated in the latter problem by James Dewar's (1842-1923) announcement of an experimental determination of the rate of helium production in pure radium salt,[63] Strutt persisted and by 1909 had obtained similar measurements for several uranium ores which corresponded very closely to the theoretical value proposed by Rutherford. This result was obviously encouraging, but the process of analyzing natural ores instead of phosphate nodules led to additional complications of its own, among them the necessity of introducing a correction for the helium produced by the controversial and little understood thorium. Moreover, the problem of finding a mineral which provided a sufficiently uniform retention of helium to be useful for dating proved extremely difficult. Strutt finally settled upon zircons, which, although geologically older than the strata in which they are found, seemed to provide a clear correlation between their helium content and their stratigraphic locations. He was consequently able to make age calculations for several samples ranging from the Paleozoic through the Tertiary.

Even with his phosphate nodules Strutt had obtained ages as great as 141 million years for the Carboniferous era. He had thus exceeded the estimates of Kelvin and many geologists while arguing that his results could at best represent a minimum. The zircon measurements pushed his estimates to 321 million years for the early Paleozoic (11 million years just for the tertiary), and by 1910 a further study of Archean sphenes had produced ages in excess of 700 million years.[64] Thus suddenly the geologists found themselves faced with an embarrassment of time almost of the order postulated by the older uniformitarians.

Boltwood was Strutt's counterpart in the early development of lead dating, but unlike the Englishman, his public pronouncements on the subject were confined to a single paper. Actually Boltwood's chief interest was in identifying the elements in the radioactive decay series. In particular, he wanted to find the final stable end products, which, he was convinced, included lead. Thus, the determination of the age of a collection of minerals from their uranium/lead ratios was merely a means toward an end: the proof that lead was the product of uranium decay. It seems likely, in fact, that Rutherford, with whom Boltwood

was then carrying on an extensive correspondence, was actually responsible for suggesting that he attempt the corollary problem of
determining the ages of his samples from their lead content.[65] In any
event, when Boltwood first outlined his problem early in 1905, his
stress was upon the importance of studying minerals of known geological ages because of the importance of time in radioactive decay.[66] But in
November of the same year he privately sent Rutherford a list of the
calculated ages of twenty-six minerals ranging up to 570 million
years.[67] During the next several months Boltwood compared the
uranium/lead ratios for numerous minerals drawing upon data from
his own analyses and published sources, and by the time he finally
published his results in 1907, he had found a consistently good correlation between these ratios and the geological ages of forty-three minerals.

Boltwood was certain that he now had convincing proof that lead
must be the end product of uranium decay, but as far as an independent calculation of geological time was concerned, many questions
remained unanswered. For example, at a very fundamental level, the
decay time of uranium was still unknown, and even the proposed
values for the much shorter decay time of radium and the equilibrium
ratio of radium and uranium were questionable. Despite these and
other uncertainties, however, Boltwood assumed values for his key
parameters based upon Rutherford's theoretical estimates — after
discounting thorium completely in the conviction that lead could not
possibly be the end product of thorium decay — and calculated the
ages of his samples. His results ranged from 410 million to an incredible 2.2 billion years. Because of the questions surrounding the rates of
radioactive decay, he naturally regarded these estimates as tentative,
but he was certain that they would take on a new significance once these
problems were solved.[68]

For all of their promise, Boltwood made no attempt to refine the
doubtful parameters in his calculations. He may, in fact, have been
warned off by his fellow countryman George F. Becker, who sharply
attacked his calculations and the excessively high results (by geological
standards) which they had produced.[69] But no doubt equally important, the exact ages of the minerals had not been Boltwood's primary
concern, and thus, once he had established the role of lead in uranium
decay to his own satisfaction, he went on to other questions. Rutherford, too, abandoned the problem except for occasional scattered
references, and no one else seriously took it up for several years.[70.]

Thus Strutt's helium method was left with no serious rival from the side of physics, and in the interval came to be widely regarded as the most elegant and possibly the most accurate measure of time available.

The hiatus lasted for nearly three years, an extraordinary period when contrasted with the frantic speed of many other developments in the early research on radioactivity. But then in 1911, one of Strutt's students, Arthur Holmes (1890-1965), revived the uranium/lead method.[71] Although initially skeptical of Boltwood's results, Holmes's examination of the two methods of radioactive dating convinced him that they could and would ultimately provide geology with the precision clocks that it required.[72] He thereupon abandoned physics for physical geology, and for the next several decades made the problem of geological time particularly his own. Utilizing the discoveries of both physics and geology, Holmes, more than anyone else, was responsible for the mature development of radioactive dating; but before he could assure the acceptance of the expanded time scale that inevitably resulted, he had first to overcome the entrenched opposition of the geologists.

Geological Reaction

From the geologists' point of view the conclusions of the physicists were interesting, even startling, but hardly convincing. In the first place, the results from helium dating and those from lead dating did not agree; and in the second, knowledge about the process of radioactive decay was still shrouded in conjecture. Even more to the point, however, geology now possessed a body of data, gathered through nearly fifty years of labor, which cried out emphatically that the physicists must be wrong. Under the circumstances, opposition was inevitable; it is only surprising that there was so little of it. Only two geologists, John Joly and George F. Becker, challenged radioactive dating directly. Two others, William J. Sollas and Frank W. Clarke lent assistance by elaborating upon and strengthening the conventional chronological calculations of geology. But most geologists simply ignored the new physical results, while a few others merely deferred to the opinions of their more outspoken fellows.[73] Silence was far from consent, however, and for several years the arguments of Joly and Becker proved substantial obstacles to the reception of the new time scales.

Of the two major opponents of radioactive dating, Joly was both the most persistent and the most influential; and with good reason. His own calculation of the earth's age from the salinity of the oceans was less than ten years old. It had gained widespread acceptance as the most accurate of the purely geological calculations of time available; it agreed well with other geological estimates; and as yet it had revealed no serious flaw. Indeed its only important challenge came from the results of radioactive dating. Even so, Joly's caution was rooted in more than the mere partisanship of intellectual paternity, although that no doubt played an understandable role in his judgments. He was, except for Holmes, the geologist best equipped to understand the arguments of the physicists and to evaluate the data and assumptions on which they were based. But whereas Holmes had come into geology only after Kelvin's arguments had been decisively challenged by radioactivity, Joly had witnessed their tyrannical influence on geology and had participated in their overthrow by the discovery of new and entirely unsuspected physical phenomena. He therefore felt fully justified in asking whether there might not be still other undiscovered phenomena which might invalidate even the new physical results. Indeed, he believed that this must be the case since the new results seemed contradicted by every geological indication of the age of the earth.

In choosing to accept the geological rather than the physical estimates of the earth's age, Joly in no way impugned the importance of radioactivity for geology. He had been among the first to recognize the geological significance of the discovery of radioactive heating, and for decades continued to defend its importance in reconciling Kelvin's conclusions with those drawn from geology. He did not seriously pursue the question, however, until after the appearance of Strutt's papers on helium dating. Then spurred by Strutt's conclusions, he began a wide-ranging study of the effect of radioactivity upon several aspects of the earth's activity. Nor did he leave the study of radioactivity *per se* entirely to the physicists, but preferred instead to become actively engaged in the physical research, including a brief collaboration with Rutherford on the sticky problem of pleochroic halos. In the end, however, when a choice had to be made between the methods and results of physics and those of geology, he unhesitatingly chose geology.

Joly's attack began gradually. Speaking before the British Association at Dublin in 1908, he made no direct reference to radioactive dating at all.[74] In fact, although he dissented from Strutt's conclusion that all the earth's radioactivity must be concentrated in a relatively thin crust, he ignored the conclusions from helium dating which Strutt published at the same time. Joly's concern was to show that the radioactivity distributed throughout the earth could account for various observed terrestrial thermal phenomena and to argue that radiothermal energy must play an important role in geological dynamics. It is nonetheless clear from several of his statements that he assumed a 100 million year old earth to be an accepted postulate, if not an established fact. A substantial part of his initial address, therefore, was given over to an elaborate deductive argument based on the production of radium from uranium, and designed to show that although present terrestrial heat losses may be radiothermal in origin, radioactive heating cannot be assumed to prolong the earth's cooling indefinitely. The heat generated by uranium and radium, he argued, cannot have been responsible for the earth's primordial molten condition unless one were willing to push the time of the *consistentior status* much further into the past than geological evidence would allow. If, on the other hand, the earth were assumed to have had an original store of heat energy from dynamic sources as Kelvin had postulated, then radiothermal energy might well have slowed down the rate of terrestrial cooling and thus might make possible the reconciliation of Kelvin's results with those of geology. This, Joly believed, was the most probable interpretation of the chronological implications of radioactivity.

He had not changed his opinion the following year when he expanded his lecture into a much more detailed monograph, but neither could he any longer ignore radioactive dating. He therefore inserted an additional chapter devoted to a two-pronged attack upon Strutt's deductions from helium dating.[75] Starting with the physical side of the problem, he conceded the principle of measuring time from helium accumulation, but questioned whether Strutt had made adequate allowance for possible errors. The questions raised were by and large the obvious ones: when and how was the original uranium introduced into the mineral; can a constant preservation of uranium and its products be assumed through time; might helium be introduced from other sources; what were the effects of helium loss; and most important, can the high degree of uniformity postulated for radioactive change really be justified? Such questions, however, were designed merely to show the weaknesses in Strutt's position. The real challenge came from geology. Clearly, Joly argued, Strutt's position was too weak to be acceptable without independent corroboration from other sources, and this corroboration simply did not exist. On the contrary, every deduction from geological evidence showed Strutt's results to be many times too high. Reviewing this evidence, Joly admitted that the traditional method of measuring depths of sediment and rates of erosion left ample room for variations and uncertainties. One need only note the results of Geikie, de Lapparent, and Sollas to get some idea of the range of variation. Nonetheless, Joly believed that calculations based solely on the bulk of accumulated sediment irrespective of its orientation eliminated many of these errors. Certainly they were not likely to be in error by as much as the factors of four or five that Strutt's results would require. The deciding objection, however, came from the evidence of solvent denudation and especially from Joly's own calculations of sodium accumulation in the oceans. The data for this calculation was, he believed, the best available, and the possible sources of error the easiest to identify and allow for. He therefore reviewed his calculations in some detail, carefully allowing for necessary corrections in his simple ratio and just as carefully explaining away the importance of unknown parameters. But though his revised result raised his original limit to about 110 million years, it still left him well below the ages postulated by Strutt. This evidence seemed to Joly to be conclusive. The individual results of the geological calculations might be in error, but it was highly improbable that they should all be in error in the same direction and by so nearly the same amount. Furthermore, given the evidence of radioactive heating, Kelvin's results could now be reconciled with geology as well. The error must therefore lie in the methods of the physicists.

In choosing to rely upon the measurement of the ocean's age, Joly assured himself of strong support. The choice was natural enough since the method had been his own creation. But it was also generally considered the most accurate of the purely geological clocks, or at least it seemed to require the smallest allowance for fluctuations in its basic parameters. Furthermore, additional data analyzed independently by Sollas and Clarke seemed clearly to vindicate Joly's original conclusions.

The details of Sollas' paper, which appeared close on the heels of Joly's book in 1909, need not be repeated.[76] It follows so closely the lines of Joly's argument that he might have written it himself. The major innovations were the incorporation of new data and the treating of the salt content of the rivers of each continent independently. The paper is nonetheless significant because it clearly demonstrates the persistence of nineteenth century geological attitudes well after the introduction of radioactivity. An early convert to even the most restrictive of Kelvin's time scales, Sollas conceded that radioactive heating might greatly extend these limitations. Beyond that, however, he made no concessions to the new science. He ignored lead dating completely and dismissed helium dating in a sentence as promising but unproven. His aim was not to isolate geology from physics, however. His message was that geologists must seek a greater precision in their science through the use of more and improved methods of measurement, and to this end they should make full use of her sister sciences:

> The only approach to Geology lies through the other sciences and endeavors to enter in some other way are responsible for much of the pseudoscience that has been perpetrated in its name.[77]

In the end, however, he took only passing notice of the latter part of his own advice, but fell back heavily upon traditional geological methods. Yet there was a change. Always before, Sollas had tried to reduce the geological estimates of time, including Joly's, to conform with Kelvin's limits. Now, from his new data he was willing to entertain estimates as high as 150 million years for the age of the oceans, although he still rejected the highest result the data would allow, 421 million years. Like Joly he was willing to accept from radioactivity a few tens of millions of years added to Kelvin's limits, but no more.

A still more detailed analysis of the sea salt argument appeared the following year.[78] The author was Frank Wigglesworth Clarke (1847-1931), the chief chemist of the United States Geological Survey and a leading figure in the development of chemistry in America. For nearly three decades, Clarke had applied the techniques of chemical analysis to the problems of geology, and his professed aim in taking up the problem of chemical denudation was to stimulate further research rather than to solve the problem of the earth's age. Nonetheless, the pioneering work of Reade and Joly could not be ignored, and with it came the question of geological time. Furthermore, it was only natural that Clarke use the newer, more detailed, and presumably more accurate information at his disposal to check the accuracy of the earlier estimates of time.

Clarke's catalogue of data included an impressive array of measurements of all the important soluble minerals in dozens of rivers all over the world. Both in extent and variety it greatly surpassed the data used by either Sollas or Joly, and in addition was weighted to allow for such variables as climate, topology, and geography. Yet when this formidable array of evidence was applied to the problem of determining the rate of sodium accumulation and the consequent age of the oceans, Joly's result was little changed. Indeed, Clarke's "crude quotient" was given as 80,726,000 years. And while corrections for recycled and precipitated salt might increase this result somewhat, or allowances for primitive oceanic sodium, marine erosion, and volcanic ejectamenta might reduce it, Clarke believed that all in all his original estimate was as probable as any that might be proposed.

Clarke had confined himself to the empirical task of quantitatively refining Joly's results through the introduction of new data. He left the problem of evaluating the method itself and of placing it in the context of the results from radioactivity and the older calculations to his colleague at the Geological Survey, George F. Becker (1847-1919). A man of great talent and learning — he had studied natural history with Agassiz, mathematics with Peirce, chemistry with Bunsen, and metallurgy at the Royal Academy of Mines in Berlin—Becker was well suited to the task. He had been drawn into geology through the influence of Clarence King and quite naturally was a strong proponent of the application of physics and chemistry to the problems of geology. But while he was one of the few geologists who could understand the details of the new physics of radioactivity, thirty years of geological field work and speculation had placed his allegiance firmly in the geologists' camp.

Becker had actually begun his attack upon radioactive dating early in 1908, over a year before Joly's doubts first appeared in print. Like Joly, he had been led into the problem by an investigation of the contribution of radioactivity to the earth's annual heat loss. But unlike his Irish counterpart, he was unwilling to grant radioactivity a very prominent role in the earth's thermal history. The question had arisen when, as a result of a debate over terrestrial cooling, Becker began a critical re-evaluation of the arguments of Kelvin and King.[79] Both men, he decided, had made fundamental errors. King had assumed that a completely solid earth was necessary for its crust to be tidally stable, and yet he had postulated conditions and accepted a solution which would produce perfect solidity only in the very recent past, whereas at a minimum, geology required a tidally rigid crust from the beginning of the Cambrian. Kelvin had erred in neglecting the effects of pressure at the earth's interior and in assuming a uniform terrestrial temperature at the time of the *consistentior status*. Becker believed that slight modifications in both methods could eliminate the errors and provide an accurate estimate of the earth's age. Mathematically, the function he produced was slightly more complicated than Kelvin's, but by making certain simplifying assumptions, he found that it could be solved by assuming a solution and working backwards to find a set of reasonable initial conditions. In this manner he arrived at 60 million years as the most probable time since the earth's total solidification. He then confidently asserted that in spite of the deficiencies in the data, he did not think it likely that subsequent research would alter his result by more than 5 million years.

Although Becker had made no allowance for the heat generated by radioactivity, his result was in good agreement with the most common geological estimates of time. He therefore concluded that, contrary to common practice since 1903, it was not really necessary to rely upon radiothermal energy to bring the older physical arguments into line with geology. At most, radioactivity need contribute only ten percent of the earth's annual heat loss. This conclusion was hardly compatible with the physicists' views on radioactive heat production or with the distribution of radioactivity in the earth's crust. And in Becker's mind it raised serious questions about some of the basic premises of radioactive theory itself. He therefore turned quite naturally to still another area where the conclusions of geology and physics seemed unreconcilable, the problem of radioactive dating.[80] Unlike the cases of Kelvin and King, however, when Becker reviewed the calculations of Boltwood and Strutt, he found that he could not offer definite alternatives to the assumptions and data which they had employed. Boltwood's uranium/lead method still gave results fifty times larger than the geologically acceptable values, and Strutt's uranium/helium estimates, although considerably smaller than Boltwood's, also remained excessive. Under the circumstances, Becker argued that geologists could not reasonably accept results that were neither consistent among themselves nor with the evidence of geology. Geology, he asserted, would still gladly welcome a valid radioactive approach to determining the earth's age, but "one condition of its acceptance would clearly be that it should give periods of the same order of magnitude as those indicated by purely geological data."[81] In short, Becker's argument, like Joly's, rested on geology. He made a few suggestions about where the physicists' assumptions might be in error, but did not bother to pursue them. The geological evidence in itself was sufficient.

Clarke's research on oceanic sodium further confirmed Becker's conviction and stimulated him to deliver what he hoped would be the *coup de grace* to the excesses of radioactive dating. Writing two years after his original challenge, he left no doubt of his intentions: "If geologists can now give convincing reasons for adhering to ages within a hundred million years, or even within two hundred, they may partly repay the heavy debt due by them to Kelvin and his intellectual heirs."[82] Having assumed the burden of proof for geology, however, Becker's performance turned out to be rather anti-climactic. After reviewing the numerical results of the most common stratigraphic measurements of the earth's age — all done prior to 1900 and all falling below 100 million years — he repeated almost verbatim his earlier revision of the thermal cooling arguments of Kelvin and King and its resulting 60 million year estimate of the age of the earth. The only original part of the paper was a similar attempt to revise Joly's sea salt method to allow for a possible lower limit for the ocean's age.

The essence of the revised argument was quite simple. Since sedimentary formations contain far less sodium than primitive rocks, the bulk of the sodium carried annually to the sea must come from the latter. But since through the ages sedimentary formations have gradually covered more and more of the earth's primitive crust, the rate of sodium erosion must have gradually diminished. Therefore, a simple ratio of present sodium erosion to the ocean's total accumulation cannot give a true measure of time. Instead, an allowance must be made for a higher rate of sodium accumulation in the past and a gradual asymptotic decrease in rate into the present. On this basis, Joly's most recent estimate of time, 144 million years, might be reduced to a maximum of 74 million. The three lines of inquiry, stratigraphy, terrestrial refrigeration, and oceanic sodium, had in Becker's opinion confirmed each other. And that being granted, he concluded, "it follows that radioactive minerals cannot have the great ages which have been attributed to them"[83] There, with no further reference to radioactivity, he rested his case. He had actually added little that was new to the discussion, but it was enough to halt any further attempts by Americans to develop the techniques of radioactive dating for nearly a decade.

Encouraged by the support given his calculations from oceanic sodium, Joly plunged still more deeply into the dispute over radioactive dating. Again in 1911 the initial thrust of his argument came from the age of the oceans, where he accepted the support of Sollas and Clarke as nearly conclusive. "It is most improbable that [their results] require amendment to the extent of 50 percent," he declared.[84] The conclusions drawn from the depth of the sedimentary column seemed to him equally positive, although he felt obliged to admit that the average rate of past sediment accumulation was still quite indeterminate. Certainly whatever their minor deficiencies, both lines of reasoning were remarkably consistent in the mutual agreement of their results. The two chief radioactive methods, on the other hand, were neither consistent among themselves nor with the results from geology. Indeed, acceptance of the larger of the radioactive estimates of time, those drawn from uranium/lead dating, would require a threefold increase in the rate of formation of recent strata over the average of ancient rates, and a fourteenfold increase in the rate at which rivers deposit salt in the oceans. Joly could perhaps countenance some fluctuation in the rates of geological processes, but variations of this magnitude were totally inadmissible. The error must therefore lie in the interpretation of the evidence from radioactivity.

Unfortunately for Joly's position, a review of the procedures of radioactive dating failed to reveal any discrepancies of the magnitude required. On the contrary, such basic parameters as the rates of production of helium and lead from a given quantity of uranium were known with much greater precision than either the rate of strata formation or the rate of sodium accumulation in the seas. Furthermore, those corrections, such as allowances for primitive lead and helium in uranium ores, which might reduce the apparent age of a sample mineral were more than counterbalanced by the need to allow for the loss of helium through diffusion or the loss of lead in soluble salts. Indeed, the need for both of the latter corrections had been established empirically while the former remained purely speculative. The only remaining possibility seemed to lie in questioning the uniformity of radioactive decay. Yet, here again the dilemma was obvious since, as Joly well knew, the rates of decay of radium and several other short lived radioelements had been shown convincingly to be uniform.

His solution was to focus attention upon the one point for which no evidence was available, the origin of uranium and thorium. If the uniform decay rates of the intermediate radioelements such as radium could not be denied, the cause of their uniformity could at least be questioned. And it was not at all improbable, he suggested, that the formation of an element through radioactive transmutation might prove a necessary condition for its subsequent uniform decay. Consequently, since no one had the slightest knowledge about the origin of uranium and thorium, he saw no reason to believe that they have *necessarily* always decayed at the present rates. Joly was obviously grasping at straws. When the suggestion was first made in 1911, he had not the slightest evidence for his speculations. Only the lack of counter evidence and the totally unacceptable alternative of abandoning geological uniformity made the suggestion at all tenable. It was hardly a satisfactory situation, and characteristically he continued to search for corroborative evidence. Within a year he found it, or so he hoped, in the rate of formation of pleochroic halos, the microscopic spherical discolorations found in certain minerals.

First discovered in the mid-nineteenth century, pleochroic halos had defied explanation until Joly began investigating them in the light of radioactivity. By 1907 he was able to show that they were caused by ionization induced by alpha particles emitted from minute isolated quantities of uranium or thorium.[85] Further study indicated that their most significant characteristics, their size and color, depended respectively upon the kinetic energy and the total number of alpha particles involved.[86] Moreover, since Rutherford had already shown that the energy of an alpha particle was a function of the disintegration rate of the element which emitted it, it was apparent that the size of each halo must be a unique characteristic of the radioactive elements contained in its nucleus. (The radioactive core of a pleochroic halo is commonly referred to as its nucleus and should not be confused with the nuclei of its constituent atoms.) Indeed, each halo consisted of a series of concentric spheres of fixed radii corresponding to the rates of decay of the originally trapped parent element and each of the product elements produced in the associated decay series. The color of a given halo, on the other hand, depends upon its age, the initial amount of radioactive material trapped in the nucleus, and the rate at which that material decays. Once these characteristics were discovered, Joly reasoned that if the approximate age of a halo and the size of its radioactive nucleus could be determined, its color should give some indication of the average rate of decay of the nuclear elements. Such measurements would be particularly useful, moreover, if they could be compared with artificially produced halos of known characteristics.

Although simple in outline, Joly's proposal presented several obvious problems. The ages of the halos involved could only be estimated approximately from their geological locations, and the minute size of their nuclei made it impossible to determine precisely the amount of radioactive material enclosed. Nonetheless, by creating a number of artificial halos for comparison and making the necessary *reasonable* assumptions, he was able to estimate the number of alpha particles generated in several halos. His results were gratifying. The number of alpha particles involved was much greater than the number which should have been produced by uniform decay over the assumed time span. For Joly the conclusion was obvious: either the halos must have originally contained some unknown additional source of alpha particles, or the rates of decay of uranium and thorium must have been much greater in the past than at present. It is especially significant that he did not even consider the possiblity that the additional particles might be due to a greater antiquity of the minerals than that assigned by geology, but seized immediately upon the alternative which could justify reducing the chronological results of lead and helium dating.

The following year, 1913, he and Rutherford reversed the direction of this analysis and attempted to date a group of minerals from the color of their pleochroic halos. Ironically, they found that their most reliable results came out *above* rather than below the corresponding dates from helium and lead dating. How to reconcile these conclusions with geological data remained inexplicable.[87] Undaunted, Joly continued to pursue the study of halos at irregular intervals for more than a decade, always convinced that they would support his case and demonstrate conclusively that the uranium/lead dates were excessive.[88]

After 1915 Joly stood almost alone in actively opposing radioactive dating. Other geologists continued to entertain doubts, but only he returned again and again to the geological and radiological calculations seeking proof for his contentions. His arguments changed little, and he continued to hope that the two lines of reasoning could be reconciled. But the driving force behind his determination was always his refusal to abandon belief in the accuracy of his results from the salinity of the oceans, and consequently, the only acceptable reconciliation of results had to be in the direction indicated by geology. Not only did the concurrence of the various lines of geological evidence require such a conclusion, he asserted, but if all else failed, the limits of the sun's energy — even allowing for radioactivity — demanded it. By 1925 he had abandoned simple geological uniformity to the extent of developing an elaborate synthesis relating the thermal output of the earth's radioactivity to a series of cyclic upheavals of its basaltic crust, but the new hypothesis still required only 160 to 300 million years.[89] To this small compromise he added the concession that the results from thorium/lead dating, which were consistently lower than the corresponding uranium/lead values, might prove to be valid. But it was not until 1930 that he showed any sign of actually yielding to the accumulated mass of radiological evidence, and like Kelvin he died without publicly relinquishing his convictions.[90]

The Expansion of the Time Scale

Radioactive dating clearly had not taken geology by storm. As Joseph Barrell succinctly summarized the situation in 1917:

> After one experience with the fallibility of physical argument notwithstanding its mathematical character, it would certainly be unwise for geologists to accept unreservedly the new and larger measurements of time given by radioactivity. There may be here, also, factors undetected and unsuspected which vitiate the results.[91]

And yet by the mid-teens, a change in attitude and opinions was indeed taking place. There were still few geologists who wholeheartedly embraced the new results from physics, and only Holmes, Joly, and Barrell actively pursued the task of dating minerals from their radioactive ratios. Nonetheless, the new discoveries had raised questions which required answers, and perhaps more important, they had awakened geologists to the necessity of re-evaluating their own methods and preconceptions, unfettered by the restrictions of Kelvin's opinions. Their results were illuminating. Even such staunch defenders of the *status quo* as George Becker felt compelled to admit that the earlier geological estimates were probably influenced by a desire to conform as nearly as possible to Kelvin's demands.[92] While among many younger geologists, reappraisal led to a growing skepticism and occasionally to an outright rejection of the methods and conclusions of their elders. Thus if Joly stood almost alone in opposing radioactive dating, the reason for his isolation lay as much in the changing attitude of his fellow geologists as in the growth of concrete physical evidence.

The leading spokesman and chief architect of the new geological attitude was Arthur Holmes. While still a student in Strutt's laboratory, Holmes had fallen almost exclusive heir to the task of developing the techniques of radioactive dating which Rutherford, Strutt, and Boltwood had abandoned. Unlike the older generation of pioneers, however, Holmes plunged wholeheartedly into the pursuit of geology. His physical training became his chief tool, but he worked steadily to master the geologists' craft as well. For several years the age of the earth was his chief concern, and he thoroughly reviewed the literature — physical, geological, and biological — of the preceding half-century. The result of this labor appeared in 1913 as the first really full-scale review and assessment of the various methods of measuring geological time.[93] Holmes's review was particularly significant, moreover, because he provided synopses of the arguments themselves rather than mere comparisons of numerical results; and in so doing, exposed many of the inconsistencies concealed in their apparent agreement. Such an approach could hardly be completely objective, of course, for Holmes was clearly committed to the results from radioactive dating. But at the same time, it could not help but be revealing.

Holmes's treatment of the classical physical arguments of Kelvin, Croll, and King was thorough and fair. But while it is admittedly significant that as late as 1913 he felt obliged to give their theories a full discussion, he limited his rebuttal to pointing out where their assumptions had been vitiated by more accurate data or by subsequent discoveries.[94] Much more important was his treatment of the geologists' arguments from denudation, sedimentation, and sodium accumulation. There, he not only pointed out the uncertainties and possible flaws in the available data, but attacked the key assumption of every geological approach, the doctrine of uniformity. For nearly a century this doctrine, which may be loosely defined as the belief that the nature and magnitudes of geological processes have remained essentially uniform since the earth's formation, had been the cornerstone of geological explanation. Naturally it had undergone certain modifications in definition since William Whewell first applied the term "uniformitarianism" to the steady state geology of Charles Lyell. Kelvin himself had forced the abandonment of the original corollary of unlimited time, and Louis Agassiz had demonstrated the existence of dramatically atypical conditions in the glacial epochs. But despite the occasional introduction of necessary qualifications, uniformity had proved to be a valuable assumption. Thus, although a few geologists at the end of the century believed that an over-reliance upon uniformity — particularly a confusion between uniformity of *degree* and uniformity of *kind* — had led geology into serious error,[95] the dominant view remained that the present was the best available key to the past. Holmes disagreed.

The problem as he saw it was that both the geological and physical calculations of the earth's age depended upon assuming uniformity in the rates of certain processes of change, and both could not be correct.[96] This, of course, was exactly the same conclusion that had led Joly to reject the assumption of uniform uranium decay, but it struck Holmes in just the opposite way. The uniform decay of the short lived radioelements had been repeatedly demonstrated by experiment, he argued, and there was no reason to assume that long-lived elements acted any differently. Geological uniformity, on the other hand, not only could not be verified empirically, but good evidence existed to show that the present does not represent a viable average for geological conditions in the past. Furthermore, he continued, there was no verifiable justification for the popular assumption that plutonic forces resulting from higher terrestrial temperatures had greatly accelerated geological changes in the past. A far more probable conclusion was that the present, with its larger and more elevated continental areas — both of which would tend to increase the amount of material denuded and deposited — was really the period of accelerated change. Such circumstantial evidence clearly indicated that the assumption of uniformity was not an accurate guide for geology, and in Holmes's opinion Joly's position was patently untenable. As long as the weight of evidence remained on the side of physics, he saw no justifiable reason for rejecting radioactive dating "unless it is conclusively demonstrated that the geological estimates are beyond question."[97] This was more than a simple replay of the old debate over who should assume the burden of proof, however, as is illustrated by Holmes's more aggressive pronouncement two years later: "the data on which the geological methods depend are of doubtful value; and the assumption of uniformity which underlies their interpretation is clearly far from being justified."[98]

Holmes's distrust of the geologists' methods was shared and sometimes even exceeded by several of his colleagues. Only a few years earlier (1911), for example, H.S. Shelton had declared that no method of determining the earth's age was of any value. In a brief flurry of papers he impartially attacked the methods and conclusions of Kelvin, Geikie, G. Darwin, Sollas, Joly, Strutt, and Holmes.[99] In Shelton's terms "rational ignorance" was the only tenable course in approaching the question of geological time. The balance of evidence, he declared, clearly pointed to a terrestrial age much greater than 100 million years. But if the old arguments were thus to be discounted, so must the new ones be treated with caution. A similar opinion came from the American paleontologist, William D. Matthew (1871-1930) who besides attacking the principle of uniformity suggested yet another chronological calculation, based upon the much discussed evolution of the horse, which would greatly extend the accepted duration of the Paleozoic.[100] While such comments were not numerous — there had never been many people actively pursuing the question of the earth's age at any one time — they appear frequently enough to indicate a growing spirit of professional self-criticism among many geologists.

Such criticism, moreover, was not confined to younger scientists. Indeed, perhaps the single most balanced assessment of the weaknesses of geology in 1915 came from the older generation, from the eminent Cambridge petrologist Alfred Harker (1859-1939). Harker was convinced that geology had been severely hampered by its reliance upon ill-conceived generalizations including the doctrine of uniformity. Thus he impartially condemned the older uniformitarians' demand for unlimited time and the attempts by Joly and his fellows to confine the earth's age within preconceived limits. Both groups, he believed, had been blinded by the naive assumption that the present can be used as a meaningful average for the past. What was necessary instead, was for geologists to recognize that geological processes are controlled by conditions which cause their rates to vary, that cyclic variation rather than uniformity may be the guiding principle in geology, and that the present may well be, indeed that it probably is, a period of excessive rather than average geological activity. But even this recognition would not be enough so long as the geologists continued in their other great shortcoming, the misuse of mathematics. It was on this latter issue that Harker's remarks were the most perceptive:

It is often said that figures can be made to prove anything; and certain it is that a series of arithmetical operations does sometimes serve as introduction to very strange conclusions. The fault, of course, is not in the tool, but in the hand that uses it. In the larger issues of geology especially, where the gulf to be bridged between data and conclusions is so often a wide one, ingenuity of reasoning ought surely to be accompanied by a due sense of responsibility in the handling of figures. . . .In wrestling with problems of the kind indicated, and, I must add, in reading some very fascinating speculations by geologists of high standing, I have often wished that some obliging mathematician would put forth a small manual of applied arithmetic for the guidance of workers in the descriptive sciences. There are absolutely necessary precautions to be observed when calculation is based upon data always partial and at best roughly approximate, and these precautions are too often neglected.[101]

In order for geology to become meaningfully quantitative, Harker argued that geologists were going to have to avail themselves of the methods and standards of the other sciences; and in the case of the earth's age, he favored the radioactive methods of Strutt and Holmes.

Harker's early training in mathematics and physics — he had graduated eighth wrangler in mathematics with a first class in physics in 1882 — may account for the severity of his criticism of his colleagues' methods, but there can be little doubt that he had placed his finger upon a critical source of trouble. Certainly the persistence of the conflict between physics and geology owed much to the geologists' unwillingness to compromise the results of their newly-won quantitative methods. For more than half a century, they had worked self-consciously to make geology an exact science, and they had made substantial progress along their chosen path of substituting calculation and measurement for qualitative description. Nonetheless, by the second decade of the twentieth century their quantitative data were often meagre and erratic in quality and their mathematics unsophisticated. And there were still very few among them who could statistically or mathematically evaluate the data they had amassed or the conclusions drawn from them. In spite of such shortcomings, however, indeed perhaps because of them, many geologists had shaken off their awe of the physicists' mathematics and could now stand firm in their confidence in geological measurements. Furthermore, if they did occasionally let their convictions exceed their evidence, they were far from being alone. After all, the geological problems that had attracted the physicists were usually those for which the least direct evidence was available. What is significant, therefore, is that the geologists had awakened to these problems and had begun to look at them critically. Thus, they felt more than justified in accusing the physicists of ignoring the facts of geology and in challenging them to prove their unsupported assertions.

The geologists' aggressive skepticism is well-illustrated by the address of T.H. Holand (1868-1947) at the British Association meeting of 1914. Taking the problem of the earth's interior as one of particular importance and complexity, Holland asserted:

> The intensity and quantity of polemical literature on scientific problems frequently varies inversely as the number of direct observations on which the discussions are based: the number and variety of theories concerning a subject thus often form a coefficient of our ignorance. Beyond the superficial observations, direct and indirect, made by geologists, now extending

> below about one two-hundredth of the earth's radius,
> we have to trust to the deductions of mathematicians
> for our ideas regarding the interior of the earth; and
> they have provided us successively with every permu-
> tation and combination possible of the three states of
> matter — solid, liquid, and gaseous . . . . [102]
>
> The mathematician apparently finds it just as easy to
> prove that the earth is solid throughout as to show by
> extrapolation from known physical values that it must
> be largely gaseous.[103]

It is indicative of Holland's skepticism that he concluded his critique
with the observation that for many of the problems of physical geology
the only solution will come to "he who guesses well."[104]

Holland's skepticism was no doubt more pronounced than was typical of his fellows, but his comments together with Harker's demonstrate rather clearly a growing caution and perhaps a growing sophistication among geologists with regard to the value and limitations of
mathematics. Those who felt obliged to extend the limits of geological
time to accommodate radioactivity did so cautiously, calling attention
to the "lamentably imperfect" state of their knowledge and seldom
going so far as the results of Holmes and Rutherford seemed to
demand.[105] On the physical side, too, even such a staunch defender of
radioactive dating as Soddy called for patience in deciding between the
chronologies produced by radioactivity and geology.[106] At long last
Huxley's warning about the limitations of the mathematical mill —
which was cited by both Holland and Harker — seemed to be striking
home.

For those who accepted the expanded time scale of radioactivity,
even in modified form, more was required than the mere amassing of
better quantitative data if they were to reconcile their views with the
results of the conventional geological methods. The basic postulates
of geological dating had to be re-examined, especially the postulate of
uniformity. The problem was hardly novel. Joly, Holmes, and Harker,
not to mention their predecessors in the nineteenth century, had raised
it repeatedly, and Harker had gone so far as to suggest cycles rather
than uniformity as the guiding principle for geology. But the real
turning point in geological reasoning came with the publication of
Joseph Barrell's detailed and masterly review of the problem in
1917.[107] Barrell (1869-1919), the professor of geology at Yale, was
studying the problem of isostasy when he became convinced of the

importance of cycles — or as he termed them, rhythms — in understanding geological change. Noting that the evidence for rhythmic change had been largely ignored or at best smoothed over by the advocates of uniformity, Barrell charged that in so doing they had minimized both the differences between the present and the past and the fact that the past was an era of constant geological variation. He consequently argued that rather than an overriding principle, uniformity should be regarded only as an initial condition for beginning research, a reference point against which to measure the oscillations of geological change. He made this suggestion, moreover, even though fully aware that it would undermine all of the quantitative measures of time based upon the assumed uniformity of geological processes, and that it would necessitate entirely new methods of calculation.

Barrell's avowed approach was geological, and several times he repeated warnings that the numbers and formulae of the physicists frequently concealed the uncertain nature of their assumptions. But it was also evident that he was personally more willing to give the benefit of the doubt to the advocates of radioactive dating than to the advocates of geological uniformity. This bias was clear as he carefully reviewed the work of Boltwood and Holmes in order to show the internal consistency of their results and the overall consistency of these results with geological ratios. It was clear, too, in his meticulous attack upon the objections to radioactive dating raised by Becker and Joly, and particularly in his detailed analysis of the data on which Becker had based his attack upon Boltwood. In geology his approach was more sweeping as he amassed great quantities of evidence to prove that denudation and sedimentation were pulsatory processes dependent upon the changing elevation of the continents and sea bottoms. The novelty of his view came, however, in his arguments for the complexity of the relationships. They were so complex, he asserted, that the simple averages used by the uniformitarians were quantitatively meaningless. Indeed, contrary to both the uniformitarians who would make the present the measure of the past, and the progressionists who would make the past a period of accelerated geological activity, Barrell argued that the proper interpretation of the evidence showed the present to be a period of extraordinarily rapid denudation, perhaps as much as ten or fifteen times greater than the average. Consequently, any calculations of the earth's age which depend upon the rate of denudation — and these included all of those based upon sediment accumulation and the salt content of the oceans — would have to be expanded accordingly, and would therefore be brought into good agreement with the results from radioactivity. Barrell went still further, however, and concluded that the variability of geological processes was so great that geologists must of necessity use radioactivity to establish "the magnitude of the framework into which the geological picture must be set."[108]

The strength of Barrell's argument lay less in its novelty than in its masterly synthesis and generalization from a vast body of published data. Rhythmic processes, after all, had been sought by geologists for decades. In fact, as early as 1895, G.K. Gilbert (1843-1918), one of America's most respected geological theorists, had used the apparent rhythm of alternating sedimentary layers to determine the duration of the Cretaceous period.[109] Much like Darwin in *On the Origin of Species*, however, Barrell overwhelmed his readers with the sheer weight of evidence drawn from every conceivable source. And if his discussion of the problems of radioactivity was less original than that given by Holmes, his treatment of the geological questions was far superior. Not everyone, of course, was willing to accept Barrell's contention that uniformity was neither a law of nature nor even a good approximation to it. No one, however, could ignore his array of evidence, and few could fail to appreciate the possible escape that the concept of rhythms afforded from the dilemma created by radioactive dating.

Barrell's work did not mark the end of the dichotomy between the physical and geological views on the age of the earth, but rather the end of heated controversy over the subject. More and more during the next decade, geologists came to accept at least part of the extended time period granted them by radioactivity. To be sure, there was no concentrated effort to reconcile the two views, and unquestionably many geologists remained skeptical about the larger of the physicists' estimates. But the defenders of the older conservative time scales, notably Joly, were in a decided minority.

Many problems still remained, of course. In physics, besides all of the unanswered questions raised by radioactivity itself, two of Kelvin's original arguments were still open for debate, tidal retardation and solar heat. Thanks to the continued interest in Chamberlin's planetesimal hypothesis, the related questions of tidal retardation and terrestrial rotation were kept alive; but though they were occasionally mentioned in the context of the problem of the earth's age, they contributed little to its solution.[110] The problem of the sun's heat was more perplexing. Geological time could not be greater than the sun's age, and yet the known sources of radioactive energy were no more satisfactory than Kelvin's gravitational hypothesis in accounting for the vast stores of heat demanded by the new chronologies. This problem, perhaps more than any other, served to keep Kelvin's influence alive and gave support to opponents of radioactive dating.[111] Not surprisingly, the few clues to the solution to the dilemma — Einstein's explanation of the energy-mass relationship and Jean Perrin's suggestion of hydrogen fusion — were too complex to be utilized by the geologists, or for that matter, by most physicists.[112] The most common attitude, therefore, was the simple conviction that some as yet unknown source of energy must exist which will account for the sun's heat. On the geological side, the contributions to the inquiry were still less concrete. For a time the attempts to formulate new geological clocks seems to have been held in abeyance. One finds instead a change in attitude. Having gotten past their original infatuation with their new quantitative methods, the geologists seemed ready to recognize that many of their theories, like those of the physicists, had been in Holmes's words, "shrewd guesses" which had "been gradually adjusted to the known facts, with the result that many geological doctrines are founded not on observation alone, but also in part on fundamental hypotheses."[113]

The area of greatest advance was in the study of radioactivity itself. Soddy's discovery of isotopes and Moseley's concept of atomic number combined to open the way to improved analyses of decay products. By 1917, lead of different atomic weights had been convincingly shown to be the end product of both thorium and uranium decay. Thus one of the most perplexing uncertainties in early attempts at radioactive dating was resolved. Indeed, the whole question of radioactive decay profited from a massive, concentrated research effort. Reams of quantitative data were gathered on the rates of radioactive decay and the formation of transmutation products. Admittedly, very little of this accumulated evidence was applied directly to the question of the earth's age, and for several years Holmes worked almost alone on the problem of radioactive dating. But the physicists continually provided him with better data for his calculations; and by placing the uniformity of radioactive decay rates and the identity of decay products almost beyond question, they had immeasurably strengthened the case for his estimates of geological time.

In England, as elsewhere, research on the earth's age, like many other pursuits, was retarded by the war. Except for Joly's continued studies of pleochroic halos, almost nothing was published on the subject between 1915 and 1920. But when the first major interdisciplinary discussion of the earth's age was held after the war, it was clear that opinions had not stood still during the interim. As had been true so often in the past, the forum for this discussion was the British Association which, appropriately enough, met in 1921 in Edinburgh, where nearly thirty years before Geikie had launched the renewed attack upon Kelvin's dictates. The occasion was an impressive joint session of the Geological, Zoological, Botanical, and Mathematical and Physical Science sections, and included as speakers Sollas, Strutt (by then Lord Rayleigh), J.W. Gregory, Harold Jeffereys, and A.S. Eddington.[114] The topics discussed were familiar:uniformity, terrestrial and solar heat, sedimentation and oceanic salinity, and radioactive dating. The only really new topic was Eddington's account of a study of Cepheid variable stars which showed that they possessed sources of energy far beyond that available from gravitational attraction. Nonetheless, two things emerged clearly from the discussion. The first was that nearly twenty years after being challenged by the discovery of radioactive heating, Kelvin's arguments still demanded and received thorough consideration. And the second was that despite the persistence of nineteenth century hypotheses, a change in attitude was indeed at hand. The weight of opinion at the meeting of 1921 clearly favored radioactive dating. Geological uniformity in the late nineteenth century sense was discounted, and with it the conventional time scales based upon sedimentary rates and sea salinity. An earth of the order of 1.5 billion years was generally accepted. Of the speakers present, only Sollas continued to hold out in favor of Joly's reservations, and even he showed signs of weakening. Geology would take its time from physics, he conceded, but he wanted to be sure of what the physicists were doing first before restructuring his own field of science.

A similar interdisciplinary symposium was sponsored the following year by the American Philosophical Society. The speakers were T. C. Chamberlin for geology, John M. Clarke for paleontology, Ernest W. Brown for astronomy, and William Duane for physics.[115] The problems discussed and the conclusions reached were very much the same as those of their English colleagues: uniformity in geology was no longer a tenable assumption; the present is a period of extraordinarily rapid geological change; radioactivity provides the only non-variable clocks for measuring geological time and these point to a terrestrial age of about 1.3 billion years; and finally, this estimate is compatible with both biology and astronomy. The strongest statements came from

Chamberlin, who took geology to task for its conservatism in holding to 100 million years. He discussed both the concept of uniformity and the process of dating the age of the earth from the accumulated depth of sediments, but it was clear that his chief target was Joly. The sea salinity argument, after all, postulated an initially molten earth, and that was a condition directly contrary to Chamberlin's planetesimal hypothesis. The pride of intellectual paternity aside, however, perhaps the most significant thing about the discussions as a whole, and about the less formal remarks before the British Association, was the indication of an increased awareness of the role played by assumptions in all of the existing geochronologies.

During the next several years very little direct effort was devoted to the problem of radioactive dating, but the scattered references to geological time in the literature all point to a general acceptance of an expanded time scale.[116] By 1925, only Joly continued to argue for a limit of a few hundred million years, and even he had abandoned uniformity in favor of Barrell's cycles of activity. He was also careful to distinguish "geological time," by which he meant the time represented by the sedimentary record and the duration of the oceans, from "the age of the earth," which he conceded might be many times greater.[117] Holmes, too, was almost alone, although not quite so isolated as Joly, in his continued efforts to improve the results of radioactive dating. Meanwhile the continued research of the physicists exposed the uncertainties in several key assumptions in Joly's analysis of pleochroic halos while building an ever stronger case for uniform rates of radioactive decay. Even the always problematic thorium began to yield its secrets as the chemistry of isotopes revealed that thorium-lead salts were more soluble than analogous compounds of uranium-lead. It was thus apparent that thorium-lead was more readily leached from exposed ores than uranium-lead, and so Joly's last argument, that thorium/lead ratios were consistently lower than uranium/lead ratios, could be explained away.[118] From the geological standpoint, however, the thing that made the increased chronology credible was the change in attitude brought about by abandoning naive uniformity in favor of cyclic patterns of geological change and the recognition that the present is a period of accelerated rather than average activity.

No one embraced Barrell's concept of rhythms more enthusiastically than Holmes, who saw in it the conclusive argument in the debate between uniformity in geology and uniformity in radioactivity. Although already convinced of the validity of radioactive dating, he had felt obliged in 1913 to give thorough consideration to the older physical and geological calculations as well. He felt no such obligation when the second edition of *The Age of the Earth* appeared in 1927. "The geological hour-glass methods are incapable of providing exact results," he declared, "because the assumption of uniform rates throughout the past cannot be granted."[119] As always for Holmes, radioactivity provided the one reliable "hour-glass" for measuring the earth's age, and he provided his readers with both an explanation and a defense of radioactive dating as well as a thorough summary of the discoveries and refinements made in the method. Barrell's concept of rhythms, however, had made it possible to avoid the complete rejection of the possibility that an accurate chronology might be constructed from geological evidence alone. Indeed, in Holmes's hands the analysis of the rhythmic upheavals of the earth's crust produced estimates of time which were in remarkable agreement with those drawn from radioactivity.[120] "It is one of the triumphs of geology," he said, "to have shown beyond all possibility of doubt that there has been ample time for all the slow transformations of life and scene that the earth has witnessed."[121]

The Report of the National Research Council

Holmes had done much to sum up the state of knowledge in 1927, but the *pièce de résistance* was yet to come. A year earlier the United States National Research Council, spurred by the growing interest in geophysics and the prodigious increase in information, had established a committee to prepare systematic treatises which would incorporate the new geophysical knowledge into authoritative reference works. One subcommittee of this group — composed of Adolph Knopf, A.C. Lane, Ernest W. Brown, Charles Schuchert, A. F. Kovarik, and Arthur Holmes — was charged with preparing a comprehensive survey of the state of knowledge about the age of the earth. Their report, which appeared in 1931, finally brought an end to the last significant vestiges of the nineteenth century's limits on geological time.[122]

The subcommittee's purpose was to compile and evaluate available information, not to pursue original research. Thus its report contained no startling new discoveries. Instead, each member except Lane prepared a summary of the state of the problem in his own area. It is significant, however, that although the geological, astronomical, and physical approaches were all covered, fully seventy-five percent of the final report was devoted to radioactive dating. Moreover, much of the remainder was devoted to the problems and errors of the older methods and to showing how they could be brought into line with the results from radioactivity, rather than to exploring their validity as independent chronometers. Given the membership of the committee, this outcome was not altogether surprising. Nonetheless, the overwhelming agreement about the probable accuracy of radioactive dating was still startling, particularly in view of the fact that in 1931 only *seven* samples had been dated which met all of the standards which the committee itself had established for accurate analysis.[123]

Two of the individual summaries were relatively insignificant except as final epitaphs for two lingering but no longer influential arguments from the older school. Knopf once again marshaled the arguments against Joly's calculation of the age of the oceans: namely, that it involved too many unknown factors, that it required too many oversimplified assumptions, and that it required the rivers to carry more salt in solution than they were capable of carrying. All of these points had been made before, but this time they had a lasting effect.[124] In similar fashion, Brown, whose task was to review the astronomical component of the problem, took up Kelvin's analysis of tidal friction. His argument was brief. Of the great mass of data on the tides that had accumulated since the mid-nineteenth century, he concentrated on a simple point: the fact that only in very shallow seas could the friction of the tides have any significant effect upon the earth's velocity of rotation. At the present time, he asserted, fully two-thirds of the friction of the seas is generated in the Bering Sea alone, and in the past such shallow seas have changed far too often for Kelvin's method to have any validity as a chronometer.[125]

Curiously, although Brown flatly declared that astronomy could contribute nothing to the question of the earth's age, he made no mention of one particularly important problem about which new information was rapidly becoming available, the energy of the sun. During the years immediately preceding the report, A.S. Eddington and James Jeans had begun to apply Einstein's principle of mass-energy conversion to the problem of solar heat, and had found that it could account for a virtually limitless store of energy — at least of the order of 10^{12} years. And in addition, Jeans had shown that the Laplacian nebular hypothesis could not account for the formation of the solar system without violating the conservation of momentum.[126] Thus, in a single sweep they had produced still another possible alter-

native to Kelvin's perplexing problem of solar heat, while undermining
one of the fundamental assumptions common to all of his arguments.
Brown's neglect of Jeans and Eddington may indicate that their conclusions were not yet widely appreciated, but within a decade they had
laid Kelvin's final argument to rest.

The longest and most important section of the report, the discussion
of radioactivity, fell quite naturally to Holmes.[127] Kovarick took on the
task of devising an improved formula for uranium/lead dating, one
which would allow for the presence of lead of other than radioactive
origin,[128] but it was Holmes who marshaled the evidence to prove that
such an approach was viable. Holmes took the opportunity to present
the most thorough case possible, and though little of what he wrote was
entirely new, he had never before presented the evidence and his
arguments in such detail. For example, he devoted considerable space
to summarizing the physics of radioactivity and to recounting the
progress of those discoveries which bore even incidentally upon the
development of radioactive dating. Occasionally his account was historically faulty; but in general, his discussion of such topics as isotope
determination, radioactive series, the identification of end products,
and the attempts to determine rates of radioactive decay was com-
mendable. Of particular importance, moreover, he presented his
account in language that could be appreciated both by physicists and by
geologists with little mathematical training.

Having reviewed and assessed the strengths and weaknesses of the
state of physical knowledge, Holmes turned to the task of formulating
the criteria necessary for a valid approach to radioactive dating. This
task alone required a careful evaluation of the physical, mineralogical,
and geological standards to be met by each sample mineral. Then the
data on sample minerals from all over the world — the result of nearly
thirty years of analysis by numerous investigators — had to be re-
examined in the light of these criteria. Only a handful of unaltered
uranium ore samples survived the screening. With equal care, he next
examined each of the various methods that had been proposed for
utilizing radioactivity as a chronometer and eliminated both the
thorium/lead and the helium methods because of the uncertainties
introduced by the loss of product substances. In the end Holmes
concluded that only the uranium/lead ratio was sufficiently reliable
(that is that the parameters needed for calculation were sufficiently
well known and the possibility of outside influences sufficiently minimal) to give meaningful results. It was thus from a small core remain-

ing from a vast array of data that he made his final calculations and arrived at his still tentative result:

> No more definite statement can therefore be made
> at present than that the age of the earth exceeds 1460
> million years, is probably not less than 1600 million
> years, and is probably much less than 3000 million
> years.[129]

Despite this cautious phrasing of his numerical results, Holmes had no doubts about the validity of the method he had employed. Given the constraints that he had imposed upon it, he was convinced that radioactive dating was not only a valid procedure, but that its results were the best obtainable. At the same time, however, he was equally conscious that even under the best of conditions certain critical questions still remained unanswered. For example, in 1931 the relationships, if any, between the actinium, uranium, and thorium decay series were still not fully understood. Nor was the contribution of actinium to the production of helium and lead. Thus it seemed more than probable that actinium might introduce an as yet uncorrected factor into the ratios used in helium and lead dating. Thorium, too, continued to present difficulties. The solubility of thorium-lead was known to introduce an unquantifiable possibility of error into any calculation of which it was a part. And at a more fundamental level, the disintegration constant of thorium continued to defy accurate determination. Yet thorium appeared so frequently in uranium ores that it had to be taken into consideration. Holmes, of course, was aware that he could not resolve all of the problems still remaining; he could only make allowances for them. Nonetheless, many problems, such as Joly's contention that radioactive decay could not be uniform, had been resolved, and he was confident that the other problems would be solved in time.

The persuasiveness of the case for radioactivity is well-illustrated by Schuchert's comments in the section of the report dealing with sedimentation.[130] Schuchert had begun his investigation in a spirit of cautious skepticism, willing to extend the limits of time beyond Joly's few hundred million years but stopping considerably short of the ages demanded by radioactive dating. He candidly admitted that he had not expected to find a sufficient depth of strata to satisfy the requirements of radioactivity, and was frankly surprised at the ease with which the task was accomplished. On the other hand, he was equally skeptical of the older methods of calculating the time from strata accumulation and an assumed rate of geological change. The evidence seemed far too persuasive that the assumption of uniformity could not be granted. Schuchert considered this dilemma carefully, and presented a full discussion of the sedimentary evidence. But in the end, he concluded that the only viable approach was to accept from radioactive dating an age of 500 million years for the lower Cambrian, and to adjust the stratigraphic determinations of the duration of each individual period using that as a standard. He believed that approached in this manner, radioactivity would make it possible to determine the rates of formation of the various strata, while stratigraphy, in turn, would provide an important check upon radioactive results. He further believed that stratigraphy would be invaluable in the task of extrapolating results into periods where radioactive data was inadequate, but he totally rejected the idea that the sedimentary record alone could provide a meaningful quantitative estimate of time.

Schuchert's exclamation that "One stratigrapher at least has wholly gone over into the camp of the radioactive workers!"[131] was indicative of the effect of the National Research Council Report. With its publication, the last lingering doubts about the expansion of the geological time scale were finally brought to an end. To be sure, the defenders of the older chronologies had been faltering for years, and if all the report had accomplished was to put an end to their faltering, it would deserve little more than passing mention. Of far greater importance was its effect in stimulating research. Largely as a direct result of its influence, an extraordinary amount of work was done in radioactive dating during the years immediately following 1931. By 1937 Holmes was able to compile a seven page bibliography of papers devoted to the development of lead and helium dating, all of them published during the preceding six years. He could also point to a growing body of astrophysical evidence which implied an even greater terrestrial age than the available radioactive measurements indicated.[132] In the two decades after 1931, this work and more that followed led to still greater increases in the allowable limits of the earth's age. During the same period, further discoveries concerning the nature of radioactivity and the multiplicity of radioactive elements made it possible to develop a whole range of procedures for dating the most recent as well as the most ancient geological periods. Another war interrupted the progress of this research, but even so, by 1946 it was possible for Frederick E. Zeuner to assemble the variety of procedures available into a graduated series providing appropriate methods for dating each epoch from historical times back to the formation of the earth.[133]

REFERENCES

[1]Curie P. and Laborde (1903), *Sur la chaleur dégagée*.

[2]Becquerel (1896), *Sur les radiations invisibles*.

[3]Rutherford and Soddy (1903), *Radioactive Change*, pp. 605-608.

[4]Rutherford and Barnes (1903), *Heating Effect of Radium Emanation*; Rutherford and Barnes (1904), *Heating Effect of Radium Emanation*.

[5]Rutherford (1904b), *Radio-Activity*, pp. 337-338.

[6]Eve (1939), *Rutherford*, p. 107.

[7]Rutherford (1904a), *Radiation and Emanation*, p. 657.

[8]Curie, E. (1937), *Madame Curie*, pp. 206-208; Eve (1939), *Rutherford*, pp. 108-109.

[9]Kelvin (1904), *Nature of Emanations from Radium*, p. 209.

[10]Eve (1939), *Rutherford*, p. 109; Thomson, J.J. (1936), *Recollections*, p. 420.

[11]The various letters in this controversy appeared in the *Times* of London on 9, 15, 18, 20, 21, 24, 28, and 31 August and 4 September 1906. A review of the whole controversy is given in Soddy (1906), *Recent Controversy on Radium*, with supplemental statements in Kelvin (1906), *Recent Radium Controversy*, and Rutherford (1906c), *Recent Radium Controversy*. Brief discussions appear in Thompson (1910), *Kelvin*, II:1190 and Eve (1939), *Rutherford*, pp. 140-141.

[12]Kelvin (1907), *Attempt to Explain the Radioactivity of Radium*, p. 232. (His italics)

[13]Planck (1949), *Scientific Autobiography*, pp. 33-34.

[14]Wilson, W. E. (1903), *Radium and Solar Energy*.

[15]Thompson (1910), *Kelvin*, II:777, Letter from Kelvin to Darwin, 8 Dec. 1881; Darwin, G. (1886), *Presidential Address*, pp. 515-518; Darwin, G. (1899), *The Tides*, pp. 300-301.

[16]Darwin, G. (1903), *Radio-Activity and the Age of the Sun*.

[17]Joly (1903a), *Radium and the Geological Age*.

[18]Hardy (1903), *Radium and the Cosmic Time Scale*.

[19]Strutt (1903), *Radium and the Sun's Heat*.

[20]Joly (1903b), *Radium and the Sun's Heat*.

[21]See, e.g.: Lindemann (1915), *Age of the Earth*.

[22]Strutt (1904a), *Becquerel Rays*, pp. 134-137; Rutherford (1904b), *Radio-Activity*, pp. 342-343. Rutherford also provided a popular account of his ideas for *Harpers*, an American magazine. (See:Rutherford (1905c), *Radium-the Cause of the Earth's Heat*.)

[23]See, e.g.: Darwin, G. (1905), *Presidential Address*, p. 29; and Duncan, R. K. (1905), *New Knowledge*, pp. 228-229.

[24]Joly (1908), *Uranium and Geology*, p. 365. (60% figure, p. 368.)

[25]Rutherford (1913), *Radioactive Substances*, pp. 653-656. (Quotation, p. 656.)

[26]Joly (1903a), *Radium and the Geological Age.*

[27]Elster and Geitel (1904), *Über die radioaktive substanz.*

[28]Liebenow (1904), *Die Radiummenge der Erde*; Rutherford (1904b), *Radio-Activity*, pp. 345-346; Rutherford (1905c), *Radium-the Cause of the Earth's Heat*, pp. 783-784; and Rutherford (1906b), *Radioactive Transformations*, pp. 213-216.

[29].Soddy (1904a), *Radio-Activity*, p. 189.

[30]*Ibid.*, pp. 182-189.

[31]Duncan (1905), *New Knowledge*, pp. 244-245.

[32]Strutt (1905b), *Radioactivity of Ordinary Matter*; Strutt (1906), *Distribution of Radium.*

[33]Sollas (1905), *Age of the Earth*, pp. 63-64.

[34]Reade (1906), *Radium and Radial Shrinkage.*

[35]Geikie, A. (1905b), *Landscape in History*, p. 214; Geikie, A. (1910), *Geology: Encyclopaedia Britannica*, p. 649.

[36]Darwin, G. (1905), *Presidential Address*, pp. 27-29.

[37]The antagonists in the debate were John M. Schaeberle (1853-1924), an astronomer, and two geologists, George F. Becker (1847-1919) and Alfred C. Lane (1863-1948). (See: Becker (1908a & b), *Age of a Cooling Globe*; Schaeberle (1908b), *Earth as a Heat-Radiating Planet*; Schaeberle (1908a), *Geological Climates*; Schaeberle (1908c), *Origin and Age*; and Lane, A. C. (1908), *Schaeberle, Becker, and the Cooling Earth.*)

[38]Examples of this view can be found in Joly's work starting with Joly (1903a), *Radium and the Geological Age*, and continuing through Joly (1925), *Surface-History of the Earth.*

[39]Rutherford (1905b), *Les Problèmes Actuels*, pp. 774-775.

[40]Boltwood (1907a), *Ultimate Disintegration Products*, pp. 86-87. For a discussion of Boltwood's debt to Rutherford see: Badash (1968), *Rutherford, Boltwood, and the Age of the Earth.*

[41]Rutherford and Soddy (1903), *Radioactive Change*, pp. 604-605. Rutherford stated this conclusion still more positively the following year in Rutherford (1904b), *Radio-Activity*, pp. 333-334.

[42]Soddy (1904b), *Life History of Radium.*

[43]Joly (1904b), *Source of Radium*; Joly (1904c), *Origin of Radium.*

[44]Boltwood (1904b), *Ratio of Radium to Uranium;* Strutt (1904c), *Occurence of Radium with Uranium;* Strutt (1905c). *On Radioactive Materials*, pp. 94-96.

[45]Soddy (1905a), *Origin of Radium*; Soddy (1905b), *Production of Radium from Uranium.*

[46]Boltwood (1905a), *Production of Radium*; Boltwood (1905c), *Origin of Radium*; Whetham (1904), *Life History of Radium*; Whetham (1905), *Origin of Radium.*

[47]Soddy (1909), *Production of Radium from Uranium.*

[48]Among the supporters of this view were: Strutt (1904a), *Becquerel Rays*, pp. 169-170

and Strutt (1905a), *Rate of Formation of Radium*; Rutherford and Boltwood (1905), *Relative Proportion of Radium and Uranium*; and Joly (1908), *Uranium and Geology*, p. 306.

⁴⁹Ramsay and Soddy (1903b), *Experiments in Radioactivity.*

⁵⁰Strutt (1905c), *Radioactive Minerals,* pp. 96-97; Rutherford (1906), *Radioactive Transformations,* pp. 183-186; Rutherford (1906a), *Mass and Velocity of ∝-Particles,* pp. 895-897; Rutherford and Hahn (1906), *Mass of ∝-Particles,* p. 906; Eve (1939), *Rutherford,* p. 129.

⁵¹Boltwood (1905c), *Origin of Radium,* p. 611.

⁵²Eve (1939), *Rutherford,* p. 132.

⁵³Strutt (1905c), *Radioactive Minerals,* p. 96; Strutt (1908a), *Association of Helium and Thorium;* Boltwood (1905b), *Ultimate Disintegration Products,* pp. 256-257. Also see: Badash (1964), *Early Developments in Radioactivity,* pp. 275-277.

⁵⁴Rutherford and Geiger (1908), *Charge and Nature of the ∝-Particle;* Rutherford and Royds (1908), *Nature of the ∝-Particle.*

⁵⁵Badash (1969), *Rutherford and Boltwood: Letters,* p. 109.

⁵⁶Boltwood (1907a), *Ultimate Disintegration Products, Pt. II*, pp. 87-88. Also see: Holmes (1914), *Lead and the Final Product of Thorium;* Holmes (1915b), *Radio-Active Problems in Geology;* and Holmes (1927), *Age of the Earth,* pp. 62-63.

⁵⁷Boltwood (1905b), *Ultimate Disintegration Products,* pp. 255-256. Also see: Badash (1968), *Rutherford, Boltwood, and the Age of the Earth,* p. 163.

⁵⁸Soddy (1905b), *Production of Radium,* pp. 778-779.

⁵⁹Rutherford (1906b), *Radioactive Transformations,* pp. 191-192.

⁶⁰Rutherford (1906a) *Mass and Velocity of ∝-Particles,* pp. 897-898. A somewhat more speculative commentary on these same results was included in Rutherford (1906b), *Radioactive Transformations,* pp. 186-191, 216-217.

⁶¹Strutt (1905c), *Radioactive Minerals,* p. 99.

⁶²Strutt (1908b), *Helium and Radio-Activity;* Strutt (1908c), *Leakage of Helium;* Strutt (1908f), *Helium in Geological Time.*

⁶³Dewar (1908a), *Rate of Production of Helium.* The results of a follow–up experiment were published two years later in Dewar (1910), *Long Period Determination of the Rate of Production of Helium.*

⁶⁴Strutt (1909a), *Helium in Geological Time, Pt. II;* Strutt (1909b), *Helium in Geological Time, Pt. III;* Strutt (1909c), *Minimum Age of Thorianite;* Strutt (1910a), *Helium in Geological Time, Pt. IV;* and Strutt (1910b), *Rate at Which Helium is Produced.*

⁶⁵Badash makes this suggestion in Badash (1968), *Rutherford, Boltwood, and the Age of the Earth,* p. 163. It seems borne out by a letter from Boltwood to Rutherford on 18 Nov. 1905. (See: Badash (1969), *Rutherford and Boltwood: Letters,* p. 103.) Rutherford did mention the possibility of using lead as a geological chronometer in the published version of his Sillman Lectures which were delivered at Yale in March 1905. (See: Rutherford (1906b), *Radioactive Transformations,* p. 192). Presumably Boltwood heard the suggestion at the lectures themselves.

[66]Boltwood (1905b), *Ultimate Disintegration Products*, pp. 253-255.

[67]Badash (1969), *Rutherford and Boltwood: Letters*, p. 100-105. Letter cited in fn. 65.

[68]Boltwood (1907a), *Ultimate Disintegration Products, Pt. II.*

[69]Becker (1908c), *Relations of Radioactivity to Cosmogony*, pp. 132-135.

[70]See, e.g.: Rutherford (1915), *Radioactive Problems in Geology*; and Rutherford (1929), *Origin of Actinium.*

[71]Holmes (1911a), *Association of Lead With Uranium.*

[72]Holmes (1913a), *Age of the Earth*; Holmes (1915a), *The Measurement of Geological Time.*

[73]For examples of this general disregard of radioactivity see: Upham (1906), *Geological Time*; and See, T.J.J. (1907), *Age of the Earth's Consolidation.*

[74]Joly (1908), *Uranium and Geology.*

[75]Joly (1909), *Radioactivity and Geology*, pp. 211-251.

[76]Sollas (1909), *Anniversary Address.*

[77]*Ibid.*, p. lxxxii.

[78]Clarke, F. W. (1910), *Chemical Denudation.*

[79]Becker's terrestrial cooling argument first appeared as part of the debate with J.M. Schaeberle and A.C. Lane mentioned in fn. 37. (See especially: Becker (1908a), *Age of a Cooling Globe.)*

[80]Becker (1908c), *Relations of Radioactivity to Cosmogony.*

[81]*Ibid.*, p. 133.

[82]Becker (1910c), *Age of the Earth.* (Quotation, p. 2.)

[83]*Ibid.*, p. 28.

[84]Joly (1911), *Age of the Earth.* (Quotation, p. 273.)

[85]Joly (1907b), *Pleochroic Halos.*

[86]Joly (1907c), *Pleochroic Halos*; Joly (1910), *Pleochroic Halos*; and Joly (1914), *Pleochroic Halos.* The last of these papers was delivered as a lecture at the University of Birmingham in Oct. 1912.

[87]Joly and Rutherford (1913), *Age of Pleochroic Halos.*

[88]Excluding numerous brief notes and published letters, Joly's most important later statements appeared in Joly (1915), *Birth-Time of the World*, pp. 1-29; Joly (1917), *Genesis of Pleochroic Halos*; Joly (1923), *Age of the Earth*; and Joly (1924), *Radioactivity and the Surface History*, pp. 1-40.

[89]Joly (1925), *Surface-History of the Earth*, pp. 145-155.

[90]Joly (1930), *Geological Importance of the Radioactivity.*

[91]Barrell (1917), *Rhythms and Geologic Time*, p. 821.

[92]Becker (1908c), *Relations of Radioactivity to Cosmogony*, p. 129.

[93]Holmes (1913a), *Age of the Earth.*

[94]*Ibid.,* p. 12-38, 126-136, 166-176. For another indication of the continuing concern with Kelvin's methods see: Anonymous (1913), *Dispute over the Antiquity of Man.*

[95]One influential geologist who did believe that a serious error was being made was Joseph Prestwich. (See: Prestwich (1895), *Collected Papers,* pp. 1-18.)

[96]Holmes (1913), *Age of the Earth,* pp. 166ff.

[97]*Ibid.,* p. 170.

[98]Holmes (1915b), *Radio-Active Problems in Geology,* pp. 432-433.

[99]Shelton (1911), *Modern Theories;* Shelton (1913-14a), *Age of the Earth;* Shelton (1913-14b), *Aspects of Geologic Time.*

[100]Matthew (1914), *Time Ratios.*

[101]Harker (1915), *Geology in Reaction,* p. 105.

[102]Holland (1914b), *Presidential Address,* p. 345.

[103]*Ibid.,* p. 348.

[104]*Ibid.,* p. 358.

[105]See, e.g.: Schuchert (1923), *Earth's Changing Surface.* This is one of several lectures delivered before the Sigma Xi at Yale in 1916-17.

[106]Soddy (1915b), *Radio-Active Problems in Geology.*

[107]Barrell (1917), *Rhythms and Geologic Time.*

[108]*Ibid.,* p. 751

[109]Gilbert (1895), *Sedimentary Measurement.*

[110]The relationship between the tides and the earth's rotation was discussed in some detail in Chamberlin, T. C. (1909b), *Former Rates of the Earth's Rotation* and Moulton (1909a), *Certain Relations among Possible Changes.* Chamberlin continued to pursue the problems of tides and terrestrial rotation in Chamberlin (1916), *Origin of the Earth* and Chamberlin (1928), *Two Solar Families.* Holmes gave the planetesimal hypothesis considerable weight in Holmes (1913), *Age of the Earth,* pp. 22-31, but abandoned it in Holmes (1927), *Age of the Earth,* pp. 19-20.

[111]See, e.g.: Lindemann (1915), *Age of the Earth;* and Stromeyer (1915), *Age of the Earth.* The problem of the sun's heat was also raised many times in various contexts by Rutherford, Strutt, Holmes, Joly, and others.

[112]I have found only one reference to Einstein in this connection before 1925. (See: Bosler (1914), *Modern Theories of the Sun,* pp. 156-157.)

[113]Holmes (1913), *Age of the Earth,* p. 23.

[114]A full account of this session was never published. *The British Association Report* for 1921, pp. 413-415, carries abridged accounts of the presentations by Rayleigh, Gregory, and Eddington. *Nature,* 27 Oct. 1921, *108*:279-284 carries much fuller texts of the statements of Rayleigh, Sollas, Gregory, and Jefferys.

[115]Chamberlin, T. C. (1922), *Age of the Earth-Geological;* Clarke, J.M. (1922), *Age of the*

Earth-Paleontological; Brown, E.W. (1922), *Age of the Earth-Astronomical*; Duane, W. (1922), *Age of the Earth-Radioactivity*. These papers were all reprinted in the *Smithsonian Report*, 1922, pp. 241-273.

[116]See, e.g.: Bidder (1927), *Presidential Address*, pp. 58-59; Eddington (1928), *Nature of the Physical World*, pp. 167-169; Jeffereys (1926), *On Prof. Joly's Theory*.

[117]Joly (1925), *Surface-History of the Earth*, pp. 145-156. Joly (1926), *Surface-History of the Earth*. The latter article is Joly's usual tenacious defense of his position, this time against the criticism of Jeffereys (See fn. 116).

[118]Holmes (1926a), *Estimates of Geological Time*; Holmes (1926b), *Geological Age of the Earth*.

[119]Holmes (1927), *Age of the Earth*, p. 44.

[120]Holmes's discussion of Barrell's rhythms is given *Ibid.*, pp. 43-79.

[121]*Ibid.*, pp. 9-10.

[122]National Research Council (1931), *Bulletin No. 80*.

[123]Knopf (1931a), *Age of the Earth*, pp. 5-8.

[124]Knopf (1931b), *Age of the Ocean*.

[125]Brown, E.W. (1931), *Age of the Earth from Astronomical Data*.

[126]Eddington (1928), *Nature of the Physical World*, pp. 167-178; Jeans (1929), *Universe Around Us*, pp. 165-183, 219-227; Holmes (1937), *Age of the Earth*, pp. 218-225.

[127]Holmes (1931), *Radioactivity and Geological Time*.

[128]Kovarik (1931), *Calculating the Age of Minerals*.

[129]Holmes (1931), *Radioactivity and Geological Time*, p. 454.

[130]Schuchert (1931), *Geochronology*.

[131]*Ibid.*, p.15.

[132]Holmes (1937), *Age of the Earth*, pp. 241-242, 248-254.

[133]Zeuner (1946), *Dating the Past*.

VII
Conclusion

The acceptance of radioactive dating finally put an end to the long domination of Kelvin's restrictive chronology and returned the estimates of the age of the earth to something approaching — but never equalling — the inconceivably vast ages of Lyell and Darwin. The lasting effects of Kelvin's influence, both positive and negative, were not so easily erased, however. For more than forty years his chronology and his hypotheses regarding the structure and internal dynamics of the earth had been among the most potent influences shaping the development of geophysical theory. And though those hypotheses had to be reassessed and sometimes discounted in the light of the discoveries of the twentieth century, the institutions, individuals, and hypotheses which put them to the test frequently had their roots in an intellectual environment which Kelvin had helped to shape.

During the latter half of the nineteenth century, perhaps the most significant effect of Kelvin's influence on geology was in the changes that it imposed upon the interpretation of uniformitarianism. Gone was the vagueness of Lyell's indefinite time spans. Once the principles of Kelvin's arguments were accepted, reckless drafts on the bank of time could no longer be permitted. Even after radioactivity greatly extended the limits of the earth's age, the time available was still finite, and above all it was definite and measurable. Similarly, the steady-state earth of Lyellian uniformity was abandoned and replaced by a world of progression or geological evolution. What remained of the doctrines of Lyell and Hutton was the belief that the activities of present geological forces provide the only reliable guide to past geological change — or what continental geologists and some modern historians of geology term *actualism* to distinguish it from the broader implications of *uniformitarianism*. By the end of the nineteenth century many, perhaps most, geologists were aware of the changes that had taken place in the meaning of the term uniformitarianism. Certainly they denounced the excesses of their predecessors quite as fervently as they opposed Kelvin's later, too restrictive estimates of time. Nonetheless, in England at least, they clung to the Whewellian label of uniformitarianism even though its meaning had become confusingly vague and amorphous.

Probably the major reason for the vagueness and confusion — other than the insistent use of an outmoded label — lay in the fact that well into the twentieth century no general agreement existed about which parts of Lyell's doctrine remained viable. Could one assume that geological forces have remained constant in both *kind* and *degree*, or have they remained constant in kind only while changing in intensity? Even more fundamentally, have the agents of geological change remained constant at all, or are the only true constants the natural laws that govern their actions? The whole spectrum of possibilities raised by these and other questions was entertained at one time or another. But in general most geologists who at the end of the century still called themselves uniformitarians were willing to admit the possibility that the intensities of geological actions might have fluctuated in the past, and indeed that they might have been greater in the past than in the present. Where geologists disagreed was over which forces may have changed, and under what conditions, and by how much. And where they unanimously agreed — if only by default — was in failing to consider the possibility that the present might be a period of accelerated rather than decelerated geological activity. Moreover, in what seems the most intriguing gap in their reasoning, while allowing for limited fluctuations and even a degree of secular change in the level of geological activity, few if any geologists doubted that the present represents an accurate mean for all geological time. Thus despite their qualifications and hedges, when the geologists of the eighties and nineties came to calculate the age of the earth, they made assumptions which would have done justice to the most ardent mid-century uniformitarian. Indeed, almost every calculation involved the implicit assumptions that throughout its history the earth has maintained a constant average land area and mean continental elevation; nearly constant climatic conditions (despite the periodic occurrence of glacial epochs); and an essentially unchanging balance in the size, distribution, and rate of action of rivers, seas, and oceans. Only by overlooking the possibility that the present may be a period of accelerated change, could geologists fit this essentially Lyellian steady-state earth into Kelvin's limited chronological frame.

Another change nurtured in part, but by no means exclusively, by the search for the earth's age was the increased emphasis upon quantification and measurement in geology. Regardless of how fruitful it may have been in the past, geologists during the second half of the nineteenth century were no longer content with qualitative description. Instead, they sought exact answers to what has occurred in the earth's history, how it has occurred, and when — answers which could only come from careful and extensive measurement. Admittedly, such questions had never been entirely absent from geology, but in the late nineteenth century they were actively cultivated as geologists sought to apply the knowledge and methods of physics, astronomy, and chemistry to their own branch of science. Thus a growing band of professionals joined the laboratory to the field as they measured, mapped, and classified the layers of accumulated strata; explored the ocean bottoms and continental shelves; analyzed the sedimentary content of rivers, lakes, and seas; and calculated the force and energy expenditure of the agents of geological change. None of these tasks was undertaken solely for the purpose of determining the earth's age, but no problem did more to stimulate their pursuit, or used their results more readily.

One obvious drawback of this zeal for quantification was the temptation to place too much emphasis upon the numbers themselves and to mistake measurement for truth. Certainly in the case of the age of the earth, the geologists fell easy prey to such temptation. Throughout most of the late nineteenth century few geologists' could match their skill at gathering quantitative data with an equivalent mathematical sophistication. At times, indeed, their approach to numbers was incredibly naïve. They seldom rounded numbers to meaningful significant figures and as a result recorded data and conclusions with an implied precision that their measurements could not hope to justify. They accepted the results of gross averages as exact solutions to complex problems, and their allowances for error were rudimentary if not totally ignored. The physicists and astronomers were more experienced in the use of numbers, but they too ascribed a degree of accuracy to their results that their reliance upon estimates and assumptions could not possibly justify. Yet these shortcomings were forgotten as the numerical results alone were repeated over and over again in the preambles or summaries of each new calculation. And stripped of all assumptions, qualifications, and hedges, the remarkable agreement among the numerical estimates — a far better agreement than could be accounted for by mere chance — became convincing proof of their mutual validity.

Lost in this mere repetition of numbers, of course, were the internal inconsistencies and hidden contradictions in the arguments. At a very basic level, for example, there was a repeated failure to distinguish among the numerical results between the geological problem of the time revealed by the stratigraphic record, and the quite separate physical problem of the time since the consolidation of the earth's crust. In short, results were cited as reinforcing each other because of mere numerical agreement when in fact they related to different problems and were actually contradictory. The emphasis on numerical agreement also obscured the fact that few of the results were totally independent and that many investigators had begun with the assumption that Kelvin must be approximately right and had interpreted their data accordingly. On the other hand, it should also be pointed out that the overwhelming agreement among the various estimates of time was not merely a case of the "fudge factor" at work. Despite the accelerated growth of science in the nineteenth century, every determination of the earth's age still involved key parameters that defied measurement. Thus, however important to the overall quantitative result, these parameters had to be estimated either by indirect calculation, analogy, or, in last resort, educated guesswork. It was therefore both common sense and good scientific practice to base one's assumptions upon the best evidence and the most reliable opinions available. It was also desirable that they produce reasonable results.

Such a procedure, regardless of the care with which it is carried out, inevitably introduces into the resulting calculations the unquantifiable influence of accepted authority. Kelvin, for example, relied upon the authority of Kant and Laplace and the general acceptance of the nebular hypothesis to support his sweeping assumptions regarding the origin and initial condition of the earth and sun. That authority helped to secure the support that his conclusions required. Similarly, the geologists looked to Lyell and Hutton for the necessary support for their assumptions, and later they too joined the physicists and astronomers in looking to Kelvin as well. Geikie, Croll, and Dana also provided standards against which to measure the reasonableness of assumptions and conclusions. As in the case of the repeated numerical results, however, this compounding of references served to obscure the indirect origins of the standards themselves. The numerous citations of Geikie and Croll, to name only the most common examples, entirely overlooked the fact that both men had accepted Kelvin's results *before* making their own assumptions and embarking on their own calculations. Indeed, Geikie was cited as an "independent author-ity" for nearly a quarter of a century before publishing his own, highly derivative calculation of time.

One result of the interdependence of accepted authorities was that the assumptions chosen produced a remarkable degree of agreement among results obtained from sometimes contradictory appoaches to the problem of time. General agreement was not unanimity, however, and the accepted authorities were never entirely harmonious. Equally significant, moreover, the individual biases and preconceptions of each investigator necessarily influenced his approach to problems and his choice of method, data, and authority. They thus introduced still another, often decisive, variable into the choice of assumptions and the consequent results. In retrospect the differences between the various approaches to the question of geological time may seem less striking than the relatively narrow range of divergence in the final results. At the time, however, those differences were sufficient to make con-troversy a persistent feature in the search for the age of the earth and an important influence on the way that search was pursued.

Perhaps the most significant feature of the controversies over the age of the earth was the degree to which the antagonists divided along disciplinary lines. From the time when Kelvin coupled the first announcement of his results to an attack upon uniformitarianism, until the final acceptance of radioactive dating nearly seventy years later, the question of the earth's age was never entirely free from some degree of tension between physics and geology. Even though the geologists yielded with surprisingly little reluctance to the first onslaught of Kelvin's physics, his continued attacks on uniformitarianism and his ever more restrictive limits on time kept the controversy very much alive. A similar pattern was repeated when the geologists first wel-comed radioactivity as a way out of the impasse imposed by Kelvin only to be confronted with still another chronology that they could not accept. In both cases, conservatism and the reluctance to give up accepted and hitherto productive concepts played important parts in the ensuing controversies. Equally important, however, the geologists saw their own legitimate methods of inquiry under attack and felt obliged to come to their defense. Admittedly, the disagreements over the earth's age were not always interdisciplinary. Debates raged within the separate sciences as well as among them, and other sciences besides physics and geology were often involved. The participants themselves, however, generally saw the problem as a conflict between physics and geology, and in retrospect, that view does not seem altogether unjustified.

From the historians' point of view, these controversies, often involving prominent figures and carried on from the rostrums of much publicized gatherings such as the British Association or in the pages of semi-popular journals such as *Nature*, provide interesting insights into the public image of science and the clubbish atmosphere surrounding Victorian scientific leadership. They have had much to say about the growing professionalism of science in nineteenth century Britain. At the same time, however, the prominent position of many of the antagonists can lend — indeed, it already has lent — too great a significance to polemical rhetoric at the expense of a more balanced view of the issues at stake. Certainly, this was often the case among the participants themselves. In the heat of debate the limitations imposed by their own assumptions were forgotten; internal contradictions were ignored; and cooler, more careful reevaluations of the evidence and arguments were neglected. Sometimes, indeed, the exigencies of conflict concealed the fact that the participants were fundamentally united in their desire to find the physical principles and geological processes that have shaped the earth's crust. But if the tensions thus generated sometimes produced more heat than light, their effects were by no means always negative. Both physics and geology were stimulated by the interaction of ideas and methods, and both ultimately profited from the exchange.

Turning finally to the central figure in this study, Kelvin seems at first glance an unlikely candidate to have had so profound an influence on geological thought. His knowledge of most aspects of geology was rudimentary at best. The problems he attacked were actually extensions of physical problems rather than questions raised by his interest in geology; and in the case of the age of the earth, he pointedly ignored both the evidence and the arguments put forward by geologists. Indeed, although his scientific interests were catholic, outside of physics they were often shallow; and during his most productive years, he was often unfamiliar with the most relevant literature in his own physical specialties, let alone that in the areas of his minor interests. Whatever his shortcomings, however, his great strength lay in his ability to recognize problems of seminal importance, to inspire enthusiasm, and to stimulate inquiry. These were the strengths that he brought to the problem of determining the age of the earth, and they, rather than his discarded calculations, provide the true measure of his influence.

Even Kelvin's most implacable opponents acknowledged the debt owed him for awakening geology to the importance of establishing an exact time scale. His attacks on uniformitarianism exposed the vagueness of the mid-century concept of limitless time, while the variety of his interests made it possible for him to grasp the disparate threads of evidence needed to establish a new chronology. The limitations of his geological knowledge, and indeed his cavalier indifference to it, were at times glaringly apparent, but his unchallenged mastery of the physical principles that he brought to bear demanded respect if not uncritical allegiance. Not the least of the reasons for his influence, moreover, lay in the fact that having once defined a problem, he would not let it rest unsolved or with the solution in dispute. Thus, either directly or indirectly, he stimulated new methods of inquiry, influenced the choice of assumptions necessary for calculation, provided the standard against which results were measured, and stood at the focus of the ensuing controversies. Radioactivity undermined the foundations of his calculations, but it could not destroy the lasting effects of his influence. Largely because of Kelvin's efforts, the problem of the age of the earth had been removed from the realm of vague cosmogonal speculation and made a legitimate subject of scientific inquiry.

Afterword

I have been understandably gratified by the reception this book has received since it first appeared fifteen years ago, and I thus welcomed the opportunity afforded by its reprinting to reexamine what I wrote then in the light of the relevant scholarship that has appeared since. That has proved both rewarding and troubling. A number of authors have commented lately on the narrow focus of most British and American studies on the history of geology. While praising the quality of many of the recent works, they have criticized the narrow concentration on British stratigraphy and the formative years of the science. In trying to assess how the scholarship of the past fifteen years might have affected my approach to the subject of this book, I too have found the gaps all too prominent. Thus, were I to rewrite the book now from scratch, I could include a much richer historical background detailing the state of British geology in 1850, and I could bring to it the insights of a number of new historiographic approaches and techniques, especially in dealing with the social contexts in which the controversies regarding the age of the earth took place. But, on the whole, I think my interpretations, my conclusions, and my major lines of evidence would remain pretty much unchanged.

When this book was written, the history of geology, at least as a topic studied by professional historians of science, was in its infancy. Although histories of geology written by and for geologists are as old as Charles Lyell and Archibald Geikie, two decades ago only a handful of historians had ventured into this major area of the history of science, and much of what had been written was strongly influenced by a whiggish mythology dating from the opening "historical" chapters of Lyell's *Principles of Geology*. At that time, too, historians of science were busily searching for "revolutions" in scientific thought, and the perceived triumph of Lyell's uniformitarianism seemed to fill the bill for geology. A few scholars, notably Reijer Hooykaas, Martin Rudwick, and W. F. Cannon, had begun to challenge this mythology and to question especially the notion that Lyell's uniformitarianism had had a revolutionary impact on nineteenth-century geology. For the most part, however, the history of geology remained in Lyell's shadow. Except for tracing the influence of geology on the work of Charles Darwin, few studies looked beyond the publication of Lyell's *Principles* in the 1830s and fewer still beyond the publication of the *Origin* in 1859.

The history of geology is still the youngest and least explored major branch of the history of science, but since 1975 the work of scholars such as Martin Rudwick, Roy Porter, Rachel Laudan, Mott Greene, James Secord, Philip Lawrence, and others have added depth, detail, and sophisticated critical insight to our emerging picture of both geological theory and geological practice in the first half of the nineteenth century. Laudan, Greene, and Lawrence have also sought to dispel the Anglo-centrism that has characterized most British and American studies of the history of geology. Without disputing Lyell's unquestioned importance and influence, these scholars have decisively refuted the notion of a Lyellian revolution in geology. They have broadened and deepened our knowledge of the nineteenth-century geological community, of the social and intellectual environment in which the geologists worked, and of the problems that interested them. None of these works deal directly with the question of the age of the earth, and only Greene focuses significantly on the second half of the century, but together they add greatly to our understanding of the scientific context in which the issues discussed in this book were first formulated and in which the disputes discussed here were fought out.

My perceptions of the historical background of the age-of-the-earth controversy were inevitably affected by the information and interpretations available to me. The works of Hooykaas and Rudwick saved me from too naive an acceptance of the influence of uniformitarianism on mid-century geological theory, but were I now to rewrite parts of chapters 1, 3, and 4, I would no doubt change at least some shades of emphasis. I would certainly be less inclined to speak of uniformitarianism as "clearly the dominant geological theory in England" (p. 90). Still, I do not think Kelvin was attacking a straw man when he equated "British popular geology" and uniformitarianism. Insofar as the question of geological time was considered at all, other than in regard to the relative age and sequence of the successive layers of strata, Lyell's vast unquantifiable ages were the common currency of mid-century geologists. Certainly in the restricted sense of extrapo-

lating both the kinds and intensities of present geological actions into
the remote past, uniformitarianism was indeed the dominant operational assumption of British stratigraphers. Both the attacks on "the
doctrine of uniformity" by its critics and the attempts to clarify its
meaning by its defenders attest to a common perception of its importance among late-century British geologists. But until the history of
geology in the second half of the nineteenth century begins to receive
some of the attention that historians have thus far focused almost exclusively on the first half, our knowledge of what the terms *uniformity*
and *uniformitarianism* entailed for geologists in the 1870s and 1880s,
and how closely they corresponded to the doctrine enunciated by
Lyell half a century earlier, will remain as uncertain as it did when this
book was first written.

Except for a recent lively interest in the establishment of plate tec-
tonics, the history of geophysics has generally been ignored by both
historians of geology and historians of physics. The late S. F. (for-
merly W. F.) Cannon argued persuasively that the investigation of the
physics of the earth played a seminal role in the evolution of the
physical sciences during the first half of the nineteenth century.
Viewed in the light of Cannon's argument, Kelvin's interest in the
questions of the earth's age and the physical conditions prevailing in
its interior and his application of the laws of physics to resolving such
questions were not isolated anomalies, but part of a well-established
research tradition. A few scattered studies of the work of William
Hopkins, J. D. Forbes, and G. H. Darwin indicate that it was a tradition that continued strong throughout the century.

Over the past two decades, however, the scholar whose work has
dealt most directly with the topics treated in this book has been Stephen
Brush. Almost alone, Brush has continued to investigate the development of geophysics and the more inclusive field of planetary physics in
the late nineteenth and early twentieth centuries. Long before this
book was completed, Brush's study of the role of thermodynamics in
history contained not only a succinct summary of Kelvin's attack on
uniformitarian geology and of the key geological responses, but also a
sweeping view of their intellectual contexts. Brush's seminal work no
doubt influenced the contents of this book in more ways than I realize
even now, and my inadvertent failure to acknowledge that fact in the
first edition is an oversight that has long troubled me and that I now
hope to remedy. More recently, his articles on the planetesimal hypothesis, on theories of the origin of the moon, and on the debates over

the physical nature of the earth's interior have greatly expanded the scope and chronological coverage of topics only touched upon in this book. His latest article extends the history of the quest for a quantitative measure of the age of the earth up to the present time. Although our approaches often differ, where my work does overlap that of Brush, our conclusions tend to supplement and reinforce one another.

Two biographical studies of Kelvin have appeared since 1979, and though they differ considerably in scope and emphasis, both provide valuable insights into the scientific context of Kelvin's interest in the age of the earth. Despite its title, this book is concerned less with the internal development of Kelvin's ideas and conclusions than with their subsequent influence. Readers interested in the full sweep of Kelvin's scientific interests or in a more detailed account of the thermodynamic basis of his geochronology can now turn either to Harold Sharlin's compact but informative biography or to Crosbie Smith's and Norton Wise's massive but more selectively focused biographical study. Smith and Wise especially provide the kind of depth and detail that I could not attempt in a study such as mine. In addition to copious summaries of Kelvin's major papers on thermodynamics and the ages of the earth and sun, they draw upon his unpublished notebooks and correspondence to fill in details of the development of his ideas. They also go far beyond what I have done here with their discussion of the importance of the second law of thermodynamics in Kelvin's geophysics and in his opposition to Lyell's uniformitarianism and with the importance that they attribute to his latitudinarian religious views in his opposition to both uniformitarianism and Darwinian evolution.

That the massive labors of Smith and Wise and their extensive use of archival sources did not lead to conclusions substantially different from mine is naturally a source of great satisfaction to me. That their work and that of Brush, Albritton, and the few others who have ventured into the field extend rather than refute the evidence and conclusions presented in this book provides perhaps the best justification available for its reprinting.

Joe D. Burchfield
February 1990

Bibliography

During the nineteenth century it was common practice for major articles (and occasionally lesser articles from noted authors) to be reprinted in several journals within a few months. In such cases, the bibliography that follows lists only the place of publication that I consulted in the preparation of this book unless an exception is specifically noted. The exceptions occur where I found it necessary to consult later reprints and, in the three important instances of the papers of Kelvin, G. H. Darwin, and E. Rutherford, where I found it convenient to rely upon collected works. In these cases I have attempted to identify either the first or the most prominent place of original publication as well as the location in the appropriate collection. The abbreviations used for the major collected works are:

Darwin, G., *Scientific Papers*: George Howard Darwin, *Scientific Papers*, 5 Vols. (Cambridge: Univ. Press, 1907-16).

Kelvin, *Popular Lectures*: Sir William Thomson, Baron Kelvin, *Popular Lectures and Addresses*, 3 Vols. (London: Macmillan, 1891-94).

Kelvin, *Mathematical Papers*: Sir William Thomson, Baron Kelvin, *Mathematical and Physical Papers*, 6 Vols. (Cambridge: Univ. Press, 1882-1911).

Rutherford, *Collected Papers*: Ernest Rutherford, *The Collected Papers of Lord Rutherford of Nelson*, 3 Vols. (London: Allen & Unwin, 1962-65).

All entries, whether books or articles, are arranged alphabetically by author or editor, and under each name by date of publication.

M.A., 1869. Professor Huxley's Anniversary Address to the Geological Society, *Geol. Mag.*, 6: 275-77.

Adhémar, Joseph Alphonse, 1860. *Révolutions de la Mer: Déluges Périodique*, 2nd ed. (Paris).

Anonymous, 1859. Darwin's Origin of Species, *Saturday Review* (London), 24 Dec., *8*:775-776.

————, 1860. Professor Owen on the Origin of Species, *Saturday Review* (London), 5 May, *9*: 573-574.

————, 1866. Historical Notice in Reference to the Retardation of the Earth's Velocity of Rotation. From Letter Addressed by Professor Fick in Zurich to Professor Poggendorff, *Phil. Mag.*, Ser. 4, *31*: 322.

————, 1869a. Geo-Theology, *Geol. Mag.*, *6*: 81-84.

————, 1869b. J. Phillips, *Vesuvius* and J. L. Lobley, *Mount Vesuvius, Etc.*, *Geol. Mag.*, *6*:122-25.

————, 1869c. Mathematics versus Geology, *Pall Mall Gazette*, 3 May, pp. 11-12.

————, 1869d. Geological Dynamics, *Geol. Mag.*, *6*: 472-476.

————, 1877. The Nebular Hypothesis and Modern Genesis, *Methodist Quarterly Review*, Jan., pp. 127-157.

————, 1906. The Problems of Geology, *Nature*, 29 Mar., *73*: 513.

————, 1909. The Thermodynamics of the Earth, *Nature*, 5 Aug., *81*: 152-153.

————, 1913. How Radium Affects the Dispute Over the Antiquity of Man, *Current Opinion*, *55*: 105-106.

————, 1938. Oceans Half Billion Years Old, Their Salt Tells Scientist, *Science News Letter*, 22 Jan., *33*:53.

Arago, D. Francois, 1834. On the Thermometrical State of the Terrestrial Globe, *Edin. New Philosophical Jour.*, *16*: 205-245.

Arrhenius, Svante, 1908. *Worlds in the Making*, tr. H. Borns (New York: Harper).

J. F. B., 1875. Croll's 'Climate and Time,' *Nature*, 17 June, *12*: 121-123, and 24 June, *12*: 141-144.

Badash, Lawrence, 1964. The Early Developments in Radioactivity, Unpub. Dissertation, Yale Univ., New Haven.

————, 1968. Rutherford, Boltwood, and the Age of the Earth: The Origin of Radioactive Dating Techniques, *Proc. Am. Phil. Soc.*, *112*:157-69.

————, 1969. *Rutherford and Boltwood: Letters on Radioactivity* (New Haven: Yale Univ.).

Baitsell, George A., ed., 1923. *The Evolution of the Earth and Its Inhabitants* (New Haven: Yale Univ.).

Ball, Sir Robert Stawell, 1886. Note on the Astronomical Theory of the Great Ice Age, *Nature*, 21 Oct., *34*: 607-608.

————, 1902. *The Earth's Beginning* (New York: Appleton).

Barrell, Joseph, 1917. Rhythms and the Measurement of Geologic Time, *Bull. of Geol. Soc. of Am.*, *28*: 745-904.

————, 1923. The Origin of the Earth. In Baitsell (1923) *Evolution of Earth*.

Barus, Carl, 1891. The Contraction of Molten Rock, *Am. Jour., Sci.*, *142:* 498-99.

————, 1893a. Mr. McGee and the Washington Symposium, *Science*, *22*: 22-23.

————, 1893b. Criticism of Mr. Fisher's Remarks on Rock Fusion, *Am. Jour. Sci.*, *146*: 140-41.

Becker, George F., 1893. Fisher's New Hypothesis, *Am. Jour. Sci.*, *146*: 137-139.

————, 1904. Present Problems in Geophysics, *Science*, 28 Oct., *20*: 545-56.

————, 1908a. Age of a Cooling Globe in Which the Initial Temperature Increases Directly as the Distance from the Surface, *Science*, *27*: 227-233.

————, 1908b. Age of Cooling Globe, *Science*, *27*: 392.

————, 1908c. Relations of Radioactivity to Cosmogony and Geology, *Bull. Geol. Soc. Am.*, *19*: 113-146.

————, 1910a. Halley on the Age of the Ocean, *Science*, 25 Mar., *31*: 459-461.

————, 1910b. Reflections on Joly's Method of Determining the Ocean's Age, *Science*, 1 April, *31*: 509-12.

————, 1910c. The Age of the Earth, *Smithsonian Misc. Collections*, Vol. 56, No. 6.

Becquerel, Henri, 1896. Sur les radiations invisibles émises par les corps phosphorescents, *Comptes Rendus de l'Académie des Sciences*, *122*: 501-503.

Benson, Allan L., 1927. *The Story of Geology* (New York: Farrar).

Berget, Alphonse, 1915. *The Earth: Its Life and Death*, tr. E. W. Barlow (New York and London: Putnam).

Bidder, G. P., 1927. Presidential Address, Section D., *British Assoc. Report* (Leeds), pp. 58-74.

Boltwood, Bertram B., 1904a. Relation Between Uranium and Radium in Some Minerals, *Nature*, 26 May, *70*: 80.

————, 1904b. On the Ratio of Radium to Uranium in Some Minerals, *Am. Jour. Sci.*, Ser. 4, *18*: 96-103.

————, 1905a. The Production of Radium from Uranium, *Am. Jour. Sci.,* Ser. 4, *20*: 239-44.

————, 1905b. On the Ultimate Disintegration Products of the Radio-Active Elements, *Am. Jour. Sci.,* Ser. 4, *20*: 253-67.

————, 1905c. The Origin of Radium, *Phil. Mag.,* Ser. 6, *9*: 599-613.

————, 1907a. On the Ultimate Disintegration Products of the Radioactive Elements, Part II, The Disintegration Products of Uranium, *The Am. Jour. Sci.,* Ser. 4, *23*: 77-88.

————, 1907b. Note on a New Radioactive Element, *Am. Jour. Sci.,* Ser. 4, *24*: 370-72.

————, 1907c. The Origin of Radium, *Nature*, 10 Oct., *76*: 589.

Bonney, T. G., 1893. *The Story of Our Planet* (New York: Cassell).

Bosler, Jean, 1914. Modern Theories of the Sun, *Smithsonian Report*, pp. 153-160.

Brown, Ernest W., 1922. The Age of the Earth from the Point of View of Astronomy, *Proc. Am. Phil. Soc., 61*: 283-85.

————, 1931. The Age of the Earth from Astronomical Data. In National Research Council (1931) *Bulletin No. 80* pp. 460-66.

Brown, J. N., 1910. Rate of Emission of α-Particles from Uranium and its Products, *Proc. Roy. Soc. Lon.,* Ser. A, *84:* 151-54.

Buckley, Jerome Hamilton, 1966. *The Triumph of Time, A Study of Victorian Concepts of Time, History, Progress, and Decadence* (Cambridge: Harvard Univ.).

Buffon, G. L. Le Clerc, Comte de, 1853-54. *Oeuvres complètes de Buffon*, P. Flourens, ed. (Paris), Vol. 9. *(Epochs of Nature)*.

————, 1894. *Des Epoques de la Nature*, L. Picard, ed. (Paris).

Bulman, G. W., 1912. Radium and Earth History, *Nature*, 14 Nov., *90*: 305.

Burdon-Sanderson, J., 1903. The Limits of Science, *Science, 31* July, *18:* 140-141.

Cannon, Walter F., 1960. The Uniformitarian-Catastrophist Debate, *Isis, 51:* 38-55.

Challis, James, 1863a. A Theory Of Zodiacal Light, *Phil. Mag.,* Ser. 4, *25*: 117-125 & 183-189.

————, 1863b. On the Source and Maintenance of the Sun's Heat, *Phil. Mag.,* Ser. 4, *25*: 460-67.

Chamberlin, Rollin T., 1926. The Origin and Early Stages of the Earth, *The Nature of the World and of Man*, ed. H. H. Newman (Chicago: Univ. of Chicago), pp. 31-55.

Chamberlin, Thomas Chrowder, 1897. A Group of Hypotheses Bearing on Climatic Changes, *Jour. Geol.*, *5*: 652-683.

————, 1899. Lord Kelvin's Address on the Age of the Earth As An Abode Fitted for Life, *Science*, 30 June, *9*: 889-901, and 7 July, *10*: 11-18.

————, ed., 1909a. *The Tidal and Other Problems, Contributions to Cosmogony and the Fundamental Problems of Geology* (Washington: Carnegie Inst.).

————, 1909b. The Former Rates of the Earth's Rotation and Their Bearings on its Deformation, *Tidal and Other Problems*, Chamberlin, ed. (1909a), pp. 3-59.

————, 1916. *The Origin of the Earth* (Chicago: Univ. of Chicago).

————, 1922. The Age of the Earth from the Geological Viewpoint, *Proc. Am. Phil. Soc.*, *61*: 247-271.

————, 1928. *The Two Solar Families* (Chicago: Univ. of Chicago).

————, 1965. The Method of Multiple Working Hypotheses, *Science*, *148*: 754-59. Reprinted from *Science*, 7 Feb. 1890.

Chamberlin, Thomas C. and Salisbury, Rollin D., 1906. *Geology*, 3 Vols. (New York: Henry Holt).

Chorley, R. J., Dunn, A. J., and Beckinsale, R. P., 1964. *The History of the Study of Land Forms* (London: Methuen).

Clarke, Frank Wigglesworth, 1910. A Preliminary Study of Chemical Denudation, *Smithsonian Misc. Collections*, Vol. 56, No. 5.

Clarke, John M., 1922. The Age of the Earth from the Paleontological Viewpoint, *Proc. Am. Phil. Soc.*, *61*: 272-82.

Clausius, R. J. E., 1854. Über eine veränderte Form des Zweiten Hauptsatzes der mechanischen Warmtheorie, *Pogg. Ann.*, *93*: 481-506.

Close, Maxwell H., 1878. The Extent of Geological Time, *Geol. Mag.*, Ser. 2, *5*: 450-55.

————, 1883. The Rigidity of the Earth, *Geol. Mag.*, Ser. 2, *10*: 48.

Coleman, A. P., 1926. *Ice Ages, Recent and Ancient* (New York: Macmillian).

Collier, Katharine Brownell, 1934. *Cosmogonies of Our Fathers, Some Theories of the Seventeenth and Eighteenth Centuries* (New York: Columbia Univ.).

Cook, O. F., 1904. Evolution and Physics, *Science*, 15 July, *20*: 87-91.

Cordier, P. L. A., 1827. Essai sur la température de l'intérieur de la terre, *Memoires de l'Académie des Sciences de l'Institut de France, 7:* 473-555.

Croll, James, 1864. On the Physical Cause of the Change of Climate during Geological Epochs, *Phil. Mag.*, Ser. 4, *28*: 121-137.

————, 1866a. On the Excentricity of the Earth's Orbit, *Phil. Mag.*, Ser. 4, *31*: 26-28.

————, 1866b. On the Physical Cause of the Submergence and Emergence of the Land During the Glacial Epoch, *Phil. Mag.*, Ser. 4, *31*: 301-306.

————, 1866c. On the Influence of the Tidal Wave on the Motion of the Moon, *Phil. Mag.*, Ser. 4, *32*: 107-111.

————, 1867a. On the Excentricity of the Earth's Orbit, and its Physical Relations to the Glacial Epoch, *Phil. Mag.*, Ser. 4, *33*: 119-131.

————, 1867b. On the Change in the Obliquity of the Ecliptic, Its Influence On the Climate of the Polar Regions and on the Level of the Sea, *Phil. Mag.*, Ser. 4, *33*: 426-445.

————, 1868. On Geological Time, and the Probable Date of the Glacial and the Upper Miocene Period, *Phil. Mag.*, Ser. 4, *35:* 363-384 and *36:* 141-154, 362-386.

————, 1869. On Mr. Murphy's Theory of the Cause of the Glacial Climate, *Geol. Mag.*, 6: 382-83.

————, 1871a. On a Method of Determining the Mean Thickness of the Sedimentary Rocks of the Globe, *Geol. Mag.*, 8: 97-102.

————, 1871b. Mean Thickness of the Sedimentary Rocks, *Geol. Mag.*, 8: 285-287.

————, 1871c. On the Age of the Earth as Determined from Tidal Retardation, *Nature*, 24 Aug., *4*: 323-24.

————, 1872. Kinetic Energy, *Nature*, 15 Aug., *6*: 324.

————, 1874. On the Physical Cause of the Submergence and Emergence of Land During the Glacial Period, *Geol. Mag.*, Ser. 2, *1*: 306-14 and 346-53.

————, 1875a. Climate and Time, *Nature*, 26 Aug., *12*: 329.

————, 1875b. *Climate and Time in Their Geological Relations* (New York; Appleton).

————, 1876a. [Report on Croll's paper "On the Tidal Retardation Argument for the Age of the Earth." Read to British Association in 1876], *Nature*, 28 Sept., *14*: 481.

————, 1876b. On the Transformation of Gravity, *Phil. Mag.*, Ser. 5, *2*: 241-254.

————, 1876c. Remarks on Mr. Burns Paper on the Mechanics of Glaciers, *Geol. Mag.*, Ser. 2, *3*: 361-64.

————, 1877. On the Probable Origin and Age of the Sun, *Quar. Jour. Sci.*, *7*: 307-326.

————, 1878a. Age of the Sun in Relation to Evolution, *Nature*, 10 Jan., *17*: 206-07; 21 Feb., *17*: 321; 11 April, *17*: 464-65.

————, 1878b. Cataclysmic Theories of Geological Climate, *Geol. Mag.*, Ser. 2, *5*: 390-398.

————, 1878c. On the Origin of Nebulae, *Phil. Mag.*, Ser. 5, *6*: 1-14.

————, 1879. Interglacial Periods, *Geol. Mag.*, Ser. 2, *6*: 480.

————, 1880a. Mr. Hill on the Cause of the Glacial Epoch, *Geol. Mag.*, Ser.2, *7*: 66-67.

————, 1880b. Aqueous Vapour in Relation to Perpetual Snow, *Geol. Mag.*, Ser. 2, *7*: 357-59.

————, 1883. On Some Controverted Points in Geological Climatology; A Reply to Professor Newcomb, Mr. Hill, and Others, *Phil. Mag.*, Ser. 5, *16*: 241-264.

————, 1884a. Examination of Mr. Alfred R. Wallace's Modification of the Physical Theory of Secular Changes of Climate, *Phil. Mag.*, Ser. 5, *17*: 81-111 & 367-376.

————, 1884b. Remarks on Professor Newcomb's 'Rejoinder,' *Phil. Mag.*, Ser. 5, *17*: 275-281.

————, 1886. *Discussions on Climate and Cosmology* (New York: Appleton).

————, 1889. On Prevailing Misconceptions Regarding the Evidence which We Ought to Expect of Former Glacial Periods, *Quar. Jour. Geol. Soc. Lon.*, *45*: 220-225.

————, 1890a. *The Philosophical Basis of Evolution* (London: Edward Stanford)

————, 1890b. *Stellar Evolution and its Relations to Geological Time* (New York: Appleton).

Crowther, J. G., 1935. *British Scientists of the 19th Century* (London: Routledge).

Culverwell, Edward P., 1895. A Criticism of the Astronomical Theory of the Ice Age, and of Lord Kelvin's Suggestions in Connection with a Genial Age at the Pole, *Geol. Mag.*, Ser 4, *2*: 3-13, 55-65.

Curie, Eve, 1937. *Madame Curie* (Garden City: Doubleday).

Curie, Pierre and Laborde, Albert, 1903. Sur la chaleur dégagée spontanément par les sels de radium, *Comptes Rendus de l'Académie des Sciences, 136*: 673-75.

Dana, James D., 1873. On Some Results of the Earth's Contraction from Cooling, Including a Discussion of the Origin of Mountains, *Am. Jour. Sci., 105*: 423-443.

————, 1880. *Manual of Geology*, 3rd ed. (New York: American).

Darwin, Charles, 1859. *On the Origin of Species* (London: Murray).

————, 1959. *The Origin of Species by Charles Darwin: A Variorum Text*, ed., Morse Peckham (Philadelphia: Univ. of Penn.).

Darwin, Francis, ed., 1888. *Life and Letters of Charles Darwin*, 3 Vols. (London: Murray).

Darwin, Francis and Seward, A. C., eds., 1903. *More Letters of Charles Darwin: A Record of His Work in a Series of Hitherto Unpublished Letters*, 2 Vols. (New York: Appleton).

Darwin, George Howard, 1876-77. On the Influence of Geological Changes on the Earth's Axis of Rotation (Abstract), *Proc. Roy. Soc. Lon., 25*: 328-32.

————, 1878a. Some Results of the Supposition of the Viscosity of the Earth, *Nature*, 4 July, *18*: 265-66.

————, 1878b. On Professor Haughton's Estimate of Geological Time, *Proc. Roy. Soc. Lon., 27*: 179-183. *Scientific Papers*, III: 47-50.

————, 1879a. On the Precession of a Viscous Spheroid, and on the Remote History of the Earth, *Phil. Trans. Roy. Soc.*, Part II, *170*: 447-530. *Scientific Papers*, II: 36-139.

————, 1879b. Problems Connected with the Tides of a Viscous Spheroid, *Phil. Trans. Roy. Soc.*, Part II, *170*: 539-593. *Scientific Papers*, II: 140-207.

————, 1879-80. A Tidal Theory of the Evolution of Satellites, *The Observatory, 3*: 79-84.

————, 1880. On the Secular Changes in the Elements of the Orbit of a Satellite Revolving about a Tidally Distorted Planet, *Phil. Trans. Roy. Soc.*, Part II, *171*: 713-891. *Scientific Papers*, II: 208-382.

————, 1886. Presidential Address, Section A, *British Assoc. Report* (Birmingham), pp. 511-78.

————, 1896. The Astronomical Theory of the Glacial Period, *Nature*, 2 Jan., *53*: 196-97.

————, 1899. *The Tides: And Kindred Phenomena in the Solar System* (Boston: Houghton).

————, 1903. Radio-Activity and the Age of the Sun, *Nature*, 24 Sept., *68*: 496.

————, 1905. Presidential Address, *British Assoc. Report* (South Africa), pp. 3-32.

————, 1907-16. *Scientific Papers*, 5 Vols. (Cambridge: Univ. Press).

Davies, Gordon L., 1969. *The Earth in Decay: A History of British Geomor-phology, 1578-1878* (New York: Elsevier).

Davison, Charles, 1887a. On a Method of Determining a Lower Limit to the Age of Stratified Rocks, *Geol. Mag.,* Ser. 3, *4*: 348-49.

————, 1887b. On the Distribution of Strain in the Earth's Crust resulting from Secular Cooling With Special Reference to the Growth of Continents and the Formation of Mountain-Chains, *Proc. Roy. Soc. Lon., 42*: 325-28.

————, 1895. The Age of the Earth, *Nature,* 25 April, *51*: 607-08.

————, 1896. The Relative Lengths of Post-Glacial Time in the Two Hemispheres, *Nature,* 11 June, *54*: 137-138.

Dawkins, W. Boyd, 1870a. The Geological Calculus, *Nature*, 17 Mar., *1*: 505-06.

————, 1870b. Geological Theory in Britain, *Edin. Rev., 131*: 39-64.

Dawson, J. William, 1894. *The Meeting-Place of Geology and History* (New York: Revell).

Descartes, Rene, 1824-26. *Oeuvres de Descartes,* V. Cousin, ed. (Paris), Vol 4. (*Le Monde*).

De Launay, C. E., 1868. On the Hypothesis of the Internal Fluidity of the Terrestrial Globe, *Geol. Mag., 5*: 507-511.

De Launay, L., 1906. *L'Histoire de la Terre* (Paris: Flammarion).

De Maillet, Benoît, 1968. *Telliamed*, tr. A. V. Carozzi (Urbana: Univ. of Ill.).

De Vries, Hugo, 1904. The Evidence of Evolution, *Science*, 23 Sept., *20*: 395-401.

————, 1906. *Species and Varieties: Their Origin by Mutation*, 2nd ed., ed., Daniel Trembly MacDougal (Chicago: Open Court).

————, 1909. *The Mutation Theory*, tr. J. B. Farmer and A. D. Darbi-shire, 2 Vols. (Chicago: Open Court).

Dewar, James, 1908a. The Rate of Production of Helium from Radium, *Proc. Roy. Soc. Lon.,* Ser. A., *81*: 280-286.

————, 1908b. The Rate of Production of Helium from Radium, *Nature*, 5 Nov., *79*: 28-29.

————, 1910. Long Period Determination of the Rate of Production of Helium from Radium, *Proc. Roy. Soc. Lon.,* Ser. A., *83*: 404-408.

Doolittle, C. L., 1901. Some Advances Made in Astronomical Science During the Nineteenth Century, *Science*, 5 July, *14*: 1-12.

Dott, R. H., Jr., 1969. James Hutton and the Concept of a Dynamic Earth, *Toward A History of Geology*, ed., Cecil J. Schneer (Cam-bridge: M.I.T.), pp. 122-141.

Duane, William, 1922. The Radio-Active Point of View, *Proc. Am. Phil. Soc., 61*: 286-88.

Dubois, Eugene, 1901. The Amount of the Circulation of Carbonate of Lime and the Age of the Earth, *Koninklijke Akademie Van Wetenschappen, Amsterdam, Proceedings of Sections of Sciences, 3*: 116-130.

Duncan, P. Martin, 1877. Presidential Address, *Quart. Jour. Geol. Soc. Lon., 33*: 64-88.

Duncan, Robert Kennedy, 1905. *The New Knowledge* (New York: Barnes).

Dutton, C. E., 1876. Critical Observations on Theories of the Earth's Physical Evolution, *Geol. Mag.*, Ser. 2, *3*: 322-28.

Dyson, F. W., 1912. The Presence of Radium in the Chromosphere, *Astronomische Nachrichten, 192*: 81-82.

Eddington, Arthur S., 1928. *The Nature of the Physical World* (Cambridge: Univ. Press).

Eiseley, Loren, 1961. *Darwin's Century: Evolution and the Men Who Discovered it* (Garden City: Doubleday-Anchor).

Ellegârd, Alvar, 1958. *Darwin and the General Reader; The Reception of Darwin's Theory of Evolution in the British Periodical Press, 1859-1872* (Goteborg).

Elster, J. and Geitel, H., 1904. Über die radioaktive Substanz deren Emanation in der Bodenluft und der Atmosphäre enhalten est, *Physikalische Zeitschrift, 5*: 11-20.

Engels, Frederick, 1940. *Dialectics of Nature*, tr. & ed., Clemens Dutt (New York: International).

Evans, John, 1876. Presidential Address, *Quar. Jour. Geol. Soc. Lon.*, 32: 53-121.

————, 1878. Address to Geological Section, British Association (Dublin), *Geol. Mag.*, Ser. 2, *5*: 411-419.

Eve, Arthur S., 1939. *Rutherford* (New York: Macmillan).

Fermor, L. L., 1913. Radio-Activity and the Age of the Earth, *Nature*, 10 July, *91*: 476-77.

Fisher, Osmond, 1867. On the Ages of 'Trail' and 'Warp,' *Geol. Mag., 4*: 193-99.

————, 1868a. Denudation, and its Agents, *Geol. Mag., 5*: 34-36.

————, 1868b. On the Elevation of Mountain Chains, with a Speculation on the Cause of Volcanic Action, *Geol. Mag., 5*: 493-95.

————, 1868c. On the Denudations of Norfolk, *Geol. Mag., 5*: 544-558.

————, 1870. Contractions of Rocks in Cooling; Reply to D. Forbes, *Geol. Mag.*, 7: 58-59.

————, 1873. On the Formation of Mountains, with a Critique on Captain Hutton's Lecture, *Geol. Mag.*, 10: 248-261.

————, 1874a. On the Formation of Mountains viewed in Connection with the Secular Cooling of the Earth, *Geol. Mag.*, Ser. 2, 1: 60-63.

————, 1874b. On the Formation of Mountains; Being a Reply to Captain Hutton's Article in the January Number, *Geol. Mag.*, Ser. 2, 1: 64-66.

————, 1875. Uniformity and 'Vulcanicity,' *Geol. Mag.*, Ser. 2, 2: 97-99.

————, 1881. *Physics of the Earth's Crust* (London: Macmillan).

————, 1882. On the Depression of Ice Loaded Lands, *Geol. Mag.*, Ser. 2, 9: 526.

————, 1883. The Rigidity (?) of the Earth, *Geol. Mag.*, Ser. 2, 10: 94-95.

————, 1893. Rigidity not to be Relied upon in Estimating the Earth's Age, *Am. Jour. Sci.*, Ser. 3, 45: 464-468.

————, 1895. On the Age of the World as Depending on the Condition of the Interior, *Geol. Mag.*, Ser.4, 2: 244-246.

————, 1900. John Joly's 'An Estimate of the Geological Age of the Earth,' *Geol. Mag.*, Ser. 4, 7: 124-132.

Forbes, David, 1867. On the Chemistry of the Primeval Earth, *Geol. Mag.*, 4: 433-44.

————, 1871. The Nature of the Earth's Interior, *Geol. Mag.*, 8: 163-173.

Foster, Michael, 1899. The Growth of Science in the Nineteenth Century, *Smithsonian Report*, pp. 163-183.

Fourier, J. B. J., 1822. *Théorie analytique de la chaleur* (Paris).

————, 1827. Mémoire sur les températures du globe terrestre et des espaces planétaires, *Memories de l'Académie des Sciences de Institute de France*, 7: 569-604.

Fowler, A., 1889. Stellar Evolution, *Nature*, 27 June, 40:. 199-200.

Geikie, Archibald, 1867. Geological Time, *Geol. Mag.*, 4: 171-72.

————, 1868a. Sir Roderick Murchison and Modern Schools of Geology, *Quar. Review*, 125: 188-217.

————, 1871. On Modern Denudation, *Trans. Geol. Soc., Glasgow*, 3: 153-90. Read, 1868.

————, 1869-72. Introductory Address, *Trans. Edin. Geol. Soc., 2:* 248-67. Read, 1872.

————, [1880]. Geology, *Encyclopaedia Britannica,* 9th ed., *10:* 212-375.

————, 1882. Work in Geology, *Charles Darwin: Memorial Notices from Nature* (London: Macmillan), pp. 15-28.

————, 1892. Presidential Address, *Brit. Assoc. Report* (Edinburgh), pp. 3-26. Footnotes refer to Geikie, A. (1905b) *Landscape,* pp. 158-197.

————, 1895. Twenty-Five Years of Geological Progress in Britain, *Nature,* 14 Feb., *51:* 367-371.

————, 1899. Presidential Address, Section C, *Brit. Assoc. Report* (Dover), pp. 718-729. Footnotes refer to Geikie, A. (1905b) *Landscape,* pp. 198-233.

————, 1905a. *The Founders of Geology,* 2nd ed. (London: Macmillan).

————, 1905b. *Landscape in History and Other Essays* (London: Macmillan).

————, 1910. Geology, *Encyclopaedia Britannica,* 11th ed., *11:* 638-74.

————, 1924. *A Long Life's Work: An Autobiography* (London: Macmillan).

Geikie, James, 1873. The Antiquity of Man in Britain, *Geol. Mag., 10:* 175-179.

————, 1874. *The Great Ice Age, and its Relation to the Antiquity of Man* (New York: Appleton).

————, 1895. *The Great Ice Age and its Relation to the Antiquity of Man,* 3rd. ed. (New York: Appleton).

George, Wilma, 1964. *Biologist Philosopher: A Study of the Life of Alfred Russel Wallace* (London: Abelard-Schuman).

Giebeler, H., 1912. Spektrographische Beobachtungen der Nova Geminorum 2 am Bonner Refraktor, *Astronomische Nachrichten,* No. 4582, *191:* 393-402.

Gilbert, Grove Karl, 1895. Sedimentary Measurement of Cretaceous Time, *Jour. of Geol., 3:* 121-127.

————, 1900. Rhythms of Geologic Time, *The Popular Science Monthly, 57:* 339-353.

Gillispie, Charles Coulston, 1959. *Genesis and Geology* (New York: Harper).

Goodchild, John G., 1896. Some Geological Evidence Regarding the

Age of the Earth, *Proc. Royal Physical Society of Edinburgh, 13:* 259-308.

————, 1897-1901. On the Maintenance of the Earth's Internal Heat, *Proc. Roy. Physical Soc. of Edinburgh, 14:* 157-58.

————, 1900. Geological Time, *Trans. Geol. Soc. Glas., 11:* 267-268.

Gray, Alonzo, 1854. *Elements of Geology* (New York: Harper).

Gray, Andrew, 1908. *Lord Kelvin: An Account of His Scientific Life and Work* (London: Dent).

Gray, J. W. and Kendall, Percy F., 1892. The Cause of the Ice Age, *British Assoc. Report* (Edinburgh), pp. 708-09.

Greene, John C., 1961. *The Death of Adam* (New York: New American Lib).

Greenwood, Joseph, 1880. Eccentricity and Glacial Epochs, *Geol. Mag.,* Ser. 2, 7: 332-33.

Gregory, John W., 1898. The Plan of the Earth and its Causes, *Smithsonian Reports,* pp. 363-388.

————, 1908. Lord Kelvin's Contributions to Geology, *Trans. Geol. Soc. Glasgow, 13:* 170-186.

————, [1912]. *The Making of the Earth* (London: Williams & Northgate).

————, 1921. The Age of the Earth, *Nature,* 27 Oct., *108:* 283-84.

Haber, Francis C., 1957. Revolution in the Concept of Historical Time: A study in the Relationship between Biblical Chronology and the Rise of Modern Science, Unpub. Dissertation, Johns Hopkins Univ., Baltimore.

————, 1959a. *The Age of the World: Moses to Darwin* (Baltimore: Johns Hopkins).

————, 1959b. Fossils and Early Cosmology, *Forerunners of Darwin, 1745-1859,* Glass, Temkin, and Straus, eds. (Baltimore: Johns Hopkins), pp. 3-29.

Haeckel, Ernst, 1899. *The History of Creation,* 2 Vols., 4th ed. (London: Kegan Paul).

Halley, Edmund, 1714-16. A Short Account of the Cause of the Saltness of the Ocean, and of the Several Lakes that Emit no Rivers; with a Proposal, by Help thereof, to Discover the Age of the World, *Phil. Trans., 29:* 296-300.

Halm, J., 1902. A New Solar Theory, *Smithsonian Reports,* pp. 165-76.

Hampden, John 1880. The Earth in its Creation, its Chronology, its Physical Features, and the One Alone Portion of the Universe Adapted to Man's Occupation and Service, *Christian Jour.* (London: Guest).

Hardy, William B., 1903. Radium and the Cosmical Time Scale, *Nature,* 8 Oct., *68*: 548.

Harker, Alfred, 1915. Geology in Reaction to the Exact Sciences, With an Excursus on Geological Time, *Nature, 95*: 105-109.

Harris, Frank, 1963. *My Life and Loves,* ed., John F. Galligher (New York: Grove).

Hastie, W., ed. & tr., 1900. *Kant's Cosmogony: As in His Essay on the Retardation of the Rotation of the Earth and his Natural History and Theory of the Heavens* (Glasgow: J. Maclehose).

Haughton, Samuel, 1865. *Manual of Geology* (London: Longman).

————, 1866. On the Change of Eccentricity of the Earth's Orbit Regarded As A Cause of Change of Climate, *Phil. Mag.,* Ser. 4, *31*: 374-76.

————, 1871. *Manual of Geology,* 3rd. ed. (London: Longman).

————, 1877. Notes on Physical Geology, *Proc. Roy. Soc. Lon., 24*: 51-63 & 535-46.

————, 1878. Physical Geology, *Nature,* 4 July, *18*: 266-268.

————, 1881. New Researches on Sun-heat and Terrestrial Radiation, and on Geological Climates, Pts. I & II, *Trans. Roy. Irish Academy,* Vol. 28, pt. 6.

————, 1886. New Researches on Sunheat, Terrestrial Radiation, etc., *Royal Irish Academy, Cunningham Memoirs,* No. III, pp. 1-76 & plates.

Haughton, Walter, ed., 1966- . *The Wellesley Index to Victorian Periodicals, 1824-1900,* 2 Vols. to date (Toronto: Univ. of Toronto).

Helmholtz, Hermann von, 1856. On the Interaction of Natural Forces, *Phil. Mag.,* Ser. 4, *11*: 489-578.

————, 1881. On the Origin of the Planetary System, *Popular Lectures on Scientific Subjects,* E. Atkinson, tr., 2nd ed., 2 Vols. (London: Longman), II: 139-197.

Hennessy, Henry, 1890. On the Physical Structure of the Earth, *Smithsonian Reports,* pp. 201-219.

Herschel, John Frederick William, 1830. *A Preliminary Discourse on the Study of Natural Philosophy,* facsimile of 1830 ed. (New York: Johnson Reprint, 1966).

————, 1833-38. Letters to: C. Lyell & R. I. Murchison, *Proc. Geol. Soc. Lon., 2*: 548-52.

————, 1835. On the Astronomical Causes which may Influence Geological Phenomena, *Trans. Geol. Soc. Lon., 3*: 293-300.

————, 1849. *Outlines of Astronomy* (London: Longman).

Hill, E., 1880. Eccentricity and Glacial Epochs, *Geol. Mag.*, Ser. 2, *7*: 10-18, 190-91, 428-29.

Himstedt, F., 1904. Über die radioaktive Emanation der Wasser und Ölquellen, *Physikalische Zeitschrift, 5*: 210-13.

Hobson, Bernard, 1892-93. The Earth's Age, *Nature*, 22 Dec., *47*: 175 and 5 Jan., *47*: 226-27.

————, 1895. The Age of the Earth, *Nature*, 11 April, *51*: 558.

Holland, Thomas H., 1914a. The Earth's Crust, *Science*, 16 Oct., *40*: 533-541.

————, 1914b. Presidential Address, Section C, *British Assoc. Report* (Australia), pp. 344-358.

Holmes, Arthur, 1911a. The Association of Lead With Uranium in Rock Minerals, and Its Application to the Measurement of Geological Time, *Proc. Roy. Soc. Lon.*, Ser. A, *85*: 248-256.

————, 1911b. The Duration of Geological Time, *Nature*, 6 July, *87*: 9-10.

————, 1913a. *The Age of the Earth* (New York: Harper).

————, 1913b. Radium and the Evolution of the Earth's Crust, *Nature*, 19 June, *91*: 398.

————, 1913c. The Terrestrial Distribution of the Radio-elements, *Nature*, 7 Aug., *91*: 582-83.

————, 1914. Lead and the Final Product of Thorium, *Nature*, 2 April, *93*: 109.

————, 1915a. Radioactivity and the Measurement of Geological Time, *Proceedings of Geologists' Assoc., 26*: 289-309.

————, 1915b. Discussion on Radio-active Problems in Geology, *British Assoc. Report* (Manchester), pp. 432-34.

————, 1926a. Estimates of Geological Time, with Special Reference to Thorium Minerals and Uranium Haloes, *Phil. Mag.*, Ser. 7, *1*: 1055-1074.

————, 1926b. The Geological Age of the Earth, *Nature*, 24 April, *117*: 592-94.

————, 1927. *The Age of the Earth, An Introduction to Geological Ideas*, 2nd ed. (New York & London: Harper).

————, 1931. Radioactivity and Geological Time. In National Research Council (1931) *Bulletin 80*, pp. 124-459.

————, 1937. *The Age of the Earth*, 3rd ed. (London: Nelson).

————, 1946. An Estimate of the Age of the Earth, *Nature*, 25 May, *157*: 680-84.

————, 1949. Lead Isotopes and the Age of the Earth, *Nature*, 19 Mar., *163*: 453-56.

————, 1965. *Principles of Physical Geology* (New York: Ronald).

Hooker, Joseph D., 1868. Presidential Address, *Brit. Assoc. Report* (Norwich), pp. lviii-lxxv.

Hooykaas, Reijer, 1963. *Natural Law and Divine Miracle: The Principle of Uniformity in Geology, Biology, and Theology* (Leiden: E. J. Brill).

————, 1966. Geological Uniformitarianism and Evolution, *Archives Intern. d'Histoire des Sciences, 19*: 3-19.

Hopkins, William, 1837-42. On the State of the Interior of the Earth, First and Second Memoirs, *Phil. Trans. Abstracts, 4*: 83, 115.

————, 1839. On the Phenomena of Precession and Nutation, Assuming the Fluidity of the Interior of the Earth, *Phil. Trans. Roy. Soc.*, pp. 381-424.

————, 1852a. Presidential Address, *Quar. Jour. Geol. Soc. Lon., 8*: xxiv-lxxx.

————, 1852b. On the Causes which May have Produced Changes in the Earth's Superficial Temperature, *Quar. Jour. Geol. Soc. Lon., 8*: 56-92.

————, 1853. Presidential Address, *Quar. Jour. Geo. Soc. Lon., 9*: xxviii-xci.

Hughes, T. McK., 1872. Lyell's Principles of Geology, *Nature*, 27 June, *6*: 165-168.

Humboldt, Alexander von, 1848. *Cosmos: A Sketch and Physical Description of the Universe*, E. C. Otte, tr., 5 Vols. (London: Bohn).

Hunt, T. Sterry, 1867. The Chemistry of the Primeval Earth, *Geol. Mag., 4*: 357-69.

————, 1872. The Rigidity of the Earth and the Liquidity of Lavas, *Nature*, 11 July, *6*: 200.

————, 1880. The Chemical and Geological Relations of the Atmosphere, *Am. Jour. Sci., 119*: 349-363.

Hutton, F. W., 1873. On the Formation of Mountains, *Geol. Mag., 10* 166-174.

Hutton, James, 1788. Theory of the Earth; or an Investigation of the Laws Observable in the Composition, Dissolution and Restoration of Land upon the Globe, *Trans. Roy. Soc. Edin., 1*: 209-304.

————, 1895. *Theory of the Earth, with Proofs and Illustrations*, 3 Vols. (Edinburgh: Cadell).

Huxley, Leonard, 1900. *Life and Letters of Thomas Henry Huxley*, 2 Vols. (New York: Appleton).

————, 1918. *Life and Letters of Sir Joseph Dalton Hooker,* 2 Vols. (New York: Appleton).

Huxley, Thomas Henry, 1869. Geological Reform, *Quar. Jour. Geol. Soc. Lon., 25:* xxxviii-liii. Footnotes refer to Huxley, T. H. (1909) *Discourses,* pp. 308-342.

————, 1871. *Lay Sermons, Addresses, & Reviews* (New York: Appleton).

————, 1897a. Lectures on Evolution, *Science and Hebrew Tradition* (New York: Appleton). Delivered 1876.

————, 1897b. *Science and Christian Tradition* (New York: Appleton).

————, 1909. *Discourses, Biological and Geological Essays* (New York: Appleton).

Irons, James Campbell, 1896. *Autobiographical Sketch of James Croll with Memoir of His Life and Work* (London: Edward Stanford).

Irvine, William, 1959. *Apes, Angels, and Victorians: Darwin, Huxley, and Evolution* (New York: Meridian).

Jeans, James, 1929. *The Universe Around Us* (New York; Macmillan).

Jeffreys, Harold, 1921. The Age of the Earth, *Nature,* 27 Oct., *108:* 284.

————, 1926. On Professor Joly's Theory of Earth History, *Phil. Mag.,* Ser. 7, *1:* 923-931.

Jenkin, Fleeming, 1867. The Origin of Species, *North British Review,* June, pp. 277-318.

Jennings, Abraham G., 1900. *The Earth and the World–How Formed?* (New York: Revell).

Johnson, Frederick, 1967. Radiocarbon Dating and Archeology in North America, *Science, 155:* 165-169m

Joly, John, 1899. An Estimate of the Geological Age of the Earth, *Smithsonian Reports,* pp. 247-288.

————, 1900a. An Estimate of the Geological Age of the Earth, *Geol. Mag.,* Ser. 4, *7:* 124-132.

————, 1900b. The Geological Age of the Earth, *Geol. Mag.,* Ser. 4, *7:* 220-225.

————, 1900c. On the Geological Age of the Earth, *Brit. Assoc. Report* (Bradford), pp. 369-379.

————, 1901. The Circulation of Salt and Geological Time, *Geol. Mag.,* Ser. 4, *8:* 344-350, 504-506.

————, 1903a. Radium and the Geological Age of the Earth, *Nature,* 1 Oct., *68:* 526.

————, 1903b. Radium and the Sun's Heat, *Nature,* 15 Oct., *68:* 572.

————, 1904a. The Life History of Radium, *Nature*, 12 May, *70*: 30.

————, 1904b. The Source of Radium, *Nature*, 26 May, *70*: 80.

————, 1904c. Origin of Radium, *Nature*, 14 July, *70*: 246.

————, 1907a. Radium and Geology, *Nature*, 7 Feb., *75*: 341-42.

————, 1907b. Pleochroic Halos, *Phil. Mag.*, Ser. 6, *13*: 381-83.

————, 1907c. Pleochroic Halos, *Nature*, 10 Oct., *76*: 589.

————, 1908. Uranium and Geology, *Smithsonian Report*, pp. 355-384.

————, 1909. *Radioactivity and Geology* (London: Archibald Constable).

————, 1910. Pleochroic Halos, *Phil. Mag.*, Ser. 6, *19*: 327-30.

————, 1911. The Age of the Earth, *Smithsonian Report*, pp. 271-293.

————, 1914. Pleochroic Haloes, *Smithsonian Report*, pp. 313-327.

————, [1915]. *The Birth-Time of the World and Other Scientific Essays* (New York: Dutton).

————, 1917. The Genesis of Pleochroic Halos, *Phil. Trans. Roy. Soc.*, Sec. A, *217*: 51-79.

————, 1923. The Age of the Earth, *Scientific Monthly*, *16*: 205-216.

————, 1924. *Radioactivity and the Surface History of the Earth* (Oxford: Clarendon).

————, 1925. *The Surface-History of the Earth* (Oxford: Clarendon).

————, 1926. The Surface History of the Earth, *Phil. Mag.*, Ser. 7, *1*: 932-939.

————, 1930. The Geological Importance of the Radioactivity of Potassium, *Nature*, 20 Dec., *126*: 953.

Joly, John and Rutherford, E., 1913. *The Age of Pleochroic Halos*, *Phil. Mag.*, Ser. 6, *25*: 644-57.

Joule, James P., 1847. On the Mechanical Equivalent of Heat as Determined by the Heat Evolved by the Friction of Fluids, *Phil. Mag.*, Ser. 3, *31*: 173-176.

Judd, John W., 1870. On the Age of the Wealden, *Brit. Assoc. Report* (Liverpool), p. 77.

————, 1875. Contributions to the Study of Volcanos, *Geol. Mag.*, Ser. 2, *2*: 1-16.

————, 1888. Presidential Address, *Quar. Jour. Geol. Soc. Lon.*, *44*: 74-84.

Kant, Immanuel, 1755. *Allgemeine Naturgeschichte und Theorie des Himmels* (Konigsberg and Leipzig).

Kelvin, William Thomson, First Baron, 1852. On the Universal Tendency in Nature to the Dissipation of Mechanical Energy, *Phil. Mag.*, Ser.4, *4*: 304-306. *Mathematical Papers*, I: 511-14.

————, 1854a. On the Mechanical Energies of the Solar System, *Phil. Mag.*, Ser. 4, *8*: 409-430. *Mathematical Papers*, II: 1-27.

————, 1854b. Note on the Possible Density of the Luminiferous Medium and on the Mechanical Value of a Cubic Mile of Sunlight, *Phil. Mag.*, Ser. 4, *9*: 36-40. *Mathematical Papers*, II: 28-33.

————, 1854c. On Mechanical Antecedents of Motion, Heat, and Light, *Brit. Assoc. Report* (Liverpool), pp. 59-63. *Mathematical Papers*, II: 34-40.

————, 1855. On the Use of Observations of Terrestrial Temperature for the Investigation of Absolute Dates in Geology, *Brit. Assoc. Report* (Glasgow), pp. 18-19. *Mathematical Papers*, II: 175-77.

————, 1854-58. Origin and Transformation of Motive Power, *Royal Instit. Proc.*, *2:* 199-204. *Mathematical Papers*, II: 182-88. Read, Feb., 1856.

————, 1859a. On the Reduction of Periodical Variations of Underground Temperature, with Applications to the Edinburgh Observations, *Brit. Assoc. Report* (Aberdeen), pp. 54-56. *Mathematical Papers*, III: 291-94.

————, 1859b. Recent Investigations of M. Le Verrier on the Motion of Mercury, *Proc. Glasgow Phil. Soc.*, *4*: 263-66. *Mathematical Papers*, V: 134-37.

————, 1860. On the Variation of the Periodic Times of the Earth and Inferior Planets, Produced by Matter Falling into the Sun, *Proc. Glas. Phil. Soc.*, *4*: 272-74. *Mathematical Papers*, V: 138-140.

————, 1861a. On the Reduction of Observations of Underground Temperature; With Application to Professor Forbes' Edinburgh Observations, and the Continued Calton Hill Series, *Trans. Roy. Soc. Edin.*, *22*: 405-27. *Mathematical Papers*, III: 261-90.

————, 1861b. Physical Considerations Regarding the Possible Age of the Sun's Heat, *Brit. Assoc. Report* (Manchester), pp. 27-28. *Mathematical Papers*, V: 141-44.

————, 1862. On the Age of the Sun's Heat, *Macmillan's Mag.*, March, *5*: 288-393. *Popular Lectures*, I: 349-68.

————, 1863a. On the Secular Cooling of the Earth, *Phil. Mag.*, Ser. 4, *25*: 1-14. *Mathematical Papers*, III: 295-311. Read, 28 April 1862.

————, 1863b. On the Rigidity of the Earth, *Phil. Trans. Roy. Soc.*, *153*: 573-82. *Mathematical Papers*, III: 312-336 (with revisions).

————, 1866a. The Doctrine of Uniformity in Geology Briefly Refuted, *Proc. Roy. Soc. Edin.*, *5*: 512-13. *Popular Lectures*, II: 6-9. Read, 18 Dec. 1865.

————, 1866b. On the Physical Cause of the Submergence and Emergence of Land during the Glacial Epoch, *Phil. Mag.*, Ser. 4, *31*: 305-06. *Mathematical Papers*, V: 157-58.

————, 1866c. On the Observations and Calculations required to Find the Tidal Retardation of the Earth's Rotation, *Phil. Mag.* (Supplement), Ser. 4, *31*: 533-37. *Mathematical Papers*, III: 337-41.

————, 1866d. On the Dissipation of Energy, *Rede Lecture*, Cambridge, 23 May 1866. Reproduced in part as: On the Observations and Calculations Required to find the Tidal Retardation of the Earth's Rotation, *Popular Lectures*, II: 65-72.

————, 1869. Note on the Meteoric Theory of the Sun's Heat, *Proc. Glas. Phil. Soc.*, 7: 111-112. *Popular Lectures*, II: 127-131.

————, 1871a. On Geological Time, *Trans. Glas. Geol. Soc.*, *3*: 1-28. *Popular Lectures*, II: 10-64. Read, 27 Feb. 1868.

————, 1871b. Of Geological Dynamics, *Trans. Glas. Geol. Soc.*, *3*: 215-240. *Popular Lectures*, II: 73-127. Read, 5 April 1869.

————, 1871c. Presidential Address, *Brit. Assoc. Report* (Edinburgh), pp. lxxxiv-cv. *Popular Lectures*, II: 132-205.

————, 1872. The Rigidity of the Earth, *Nature*, 18 Jan., *5*: 223-224. *Mathematical Papers*, V: 163-170.

————, 1874. Influence of Geological Changes on the Earth's Rotation, *Proc. Glas. Geol. Soc.*, *4*: 311-13. *Mathematical Papers*, V: 171-73.

————, 1876. Review of the Evidence Regarding the Physical Condition of the Earth, *Brit. Assoc. Report* (Glasgow), pp. 1-12. *Popular Lectures*, II: 238-72.

————, 1877a. The Sun's Heat, *Good Words*, March, pp. 149-53.

————, 1877b. Geological Climate, *Trans. Glas. Geol. Soc.*, *5*: 238-50. *Popular Lectures*, II: 273-98.

————, 1878. Problems Relating to Underground Temperature: A Fragment, *Phil. Mag.*, Ser. 5, *5*: 370-74. *Mathematical Papers*, V: 175-80.

————, 1881. On the Sources of Energy in Nature Available to Man for the Production of Mechanical Effect, *Brit. Assoc. Report* (York), pp. 513-18. *Popular Lectures*, II: 433-50.

————, 1882. The Internal Condition of the Earth; As to Temperature, Fluidity, and Rigidity, *Trans. Glas. Geol. Soc.*, *6*: 38-49. *Popular Lectures, 11:* 299-318. Read, 14 Feb. 1878.

————, 1882-1911. *Mathematical and Physical Papers*, 6 Vols. (Cambridge: Univ. Press).

————, 1889. On the Sun's Heat, *Proc. Roy. Instit., 12*: 1-12. *Popular Lectures*, I: 376-429. Read, 21 Jan. 1887.

————, 1891-94. *Popular Lectures and Addresses*, 3 Vols. (London: Macmillan).

————, 1892a. On the Dissipation of Energy, *Fortnightly Review*, March, *57 (o.s.)* or *51 (n.s.):* 313-21. *Popular Lectures*, II: 451-74.

————, 1892b. Comment le Soleil à Commencé à Brûler, *L'Astronomie, 11*: 361-67. *Mathematical Papers*, V: 191-97.

————, 1895. The Age of the Earth, *Nature*, 7 March, *51*: 438-40.

————, 1899. The Age of the Earth As an Abode Fitted for Life, *Jour. Victoria Instit., 31*: 11-35. *Mathematical Papers*, V: 205-30. Read, 1897.

————, 1901. On the Clustering of Gravitational Matter in any Part of the Universe, *Brit. Assoc. Report* (Glasgow), pp. 563-68.

————, 1904. Contribution to the Discussion of the Nature of Emanations From Radium, *Phil. Mag.*, Ser. 6, *7*: 220-22. *Mathematical Papers*, VI: 206-09.

————, 1905. Plan of an Atom to be Capable of Storing an Electrion with Enormous Energy for Radio-Activity, *Phil. Mag.*, Ser. 6, *10*: 695-98. *Mathematical Papers*, VI: 227-30.

————, 1906. The Recent Radium Controversy, *Nature*, 27 Sept., *74*: 539.

————, 1907. An Attempt to Explain the Radioactivity of Radium, *Phil. Mag.*, Ser. 7, *13*: 313-16. *Mathematical Papers*, VI: 231-34.

Kelvin, Lord, and Murray, J. R. E., 1895. On the Temperature Variations of the Thermal Conductivity of Rocks, *Proc. Glas. Phil. Soc., 26*: 227-232.

Kelvin, Lord, and Tait, P. G., 1962. *Treatise on Natural Philosophy*, 2 Vols., 2nd ed. (New York: Dover). Reprinted from the last revised edition of 1902.

Kendall, Percy F., 1896. The Date of the Glacial Period, *Nature*, 6 Aug., *54*: 319.

Keppel, William Coutts, Earl of Albemarle, 1876. Modern Philosophers on the Probable Age of the World, *Quart. Rev., 142*: 202-232.

Keyes, Charles Rollin, 1906. The Age of Our Earth, *Review of Reviews, 33*: 467-68.

King, Agnes Gardner, 1925. *Kelvin the Man* (London: Hodder & Stoughton).

King, Clarence, 1877. Catastrophism and Evolution, *The American Naturalist, 11*: 449-70.

——————, 1893. The Age of the Earth, *Am. Jour. Sci., 145*: 1-20.

King, Elizabeth, 1909. *Lord Kelvin's Early Home* (London: Macmillan).

Knopf, Adolph, 1931a. The Age of the Earth, Summary of Principal Results. In National Research Council (1931), *Bulletin No. 80,* pp. 3-9.

——————, 1931b. Age of the Ocean. In National Research Council (1931), *Bulletin No. 80,* pp. 65-71.

Knott, Cargill Gilston, 1911. *Life and Scientific Work of Peter Guthrie Tait* (Cambridge: Univ. Press).

Koenigsberger, J., 1906. Über den Temperaturgradienten der Erde bei Annahme radioaktiver und chemischer Prozesse, *Physikalische Zeitschrift, 7*: 297-300.

Kovarik, Alois F., 1931. Calculating the Age of Minerals from Radioactivity Data and Principles. In National Research Council (1931) *Bulletin No. 80,* pp. 73-123.

Lamarck, J. B., 1964. *Hydrogeology,* tr. Albert V. Carozzi (Urbana: Univ. of Ill.).

Lane, Alfred C., 1908. Schaeberle, Becker and the Cooling Earth, *Science,* 10 April, *27*: 589-592.

Lane, J. Homer, 1870. On the Theoretical Temperature of the Sun; Under the Hypothesis of a Gaseous Mass Maintaining its Volume by its Internal Heat, and Depending on Laws of Gases as Known to Terrestrial Experiment, *Am. Jour. Sci., 10Q*: 57-74.

Lankester, E. Ray, 1903. The Limits of Science, *Science,* 31 July, *27*: 143-145.

Laplace, Pierre Simon, 1796. *Exposition du Système du Monde* (Paris).

Lapparent, A. de, 1889-90. De la mesure du temps par les phénomènes de sédimentation, *Bulletin de la Société Géologique de France,* Ser. 3, *18*: 351-355.

Lapworth, Charles, 1903. The Relations of Geology, *Smithsonian Reports,* pp. 363-390.

[Larmor, Joseph], 1908. William Thomson, Baron Kelvin of Largs, 1824-1907, *Proc. Roy. Soc. Lon.,* Ser. A, *81*.

Le Conte, Joseph, 1873. On the Formation of the Features of the Earth-surface. Reply to the Criticisms of T. Sterry Hunt, *Am. Jour. Sci., 105*: 448-53.

————, 1884. *Elements of Geology*, Revised ed. (New York: Appleton).

————, 1889. The General Interior Condition of the Earth, *The American Geologist, 4*: 38-44.

————, 1900. A Century of Geology, *Smithsonian Report*, pp. 265-287.

Leibniz, G. W., 1859. *Protogée ou de la formation et des révolutions du globe*, B. de Saint Germain (Paris). First published in 1749.

Liebenow, C., 1904. Notiz über die Radiummenge der Erde, *Physikalische Zeitschrift, 5*: 625-626.

Lindemann, F. A., 1915. The Age of the Earth, *Nature*, 22 April *95*: 203-204.

Litchfield, Henrietta, ed., 1915. *Emma Darwin: A Century of Family Letters*, 2 Vols. (London: Murray).

Lodge, Oliver J., 1894. Mathematical Geology, *Nature*, 26 July, *50*: 289-93.

————, 1903. The Limits of Science, *Science*, 31 July, *27*: 145-46.

Lyell, Charles, 1830-33. *Principles of Geology*, 3 Vols. (London: Murray).

————, 1850a. Presidential Address, *Quar. Jour. Geol. Soc. Lon., 6*: xxxii-lxv.

————, 1850b. *Principles of Geology*, 8th ed. (London: Murray).

————, 1851. Presidential Address, *Quar. Jour. Geol. Soc. Lon., 7*: xxxii-lxxv.

————, 1854. *Principles of Geology*, 9th ed., (New York; Appleton).

————, 1867-68. *Principles of Geology*, 10th ed., 2 Vols. (London: Murray).

————, 1877. *Principles of Geology*, 11th ed., 2 Vols. (New York: Appleton).

Lyell, [Kathrine], 1881. *Life, Letters, and Journals of Sir Charles Lyell, Bart.*, 2 Vols. (London: Murray).

McFarland, R. W., 1880. Perihelion and Eccentricity, *Am. Jour. Sci., 120*: 105-11.

McGee, William J., 1879. Notes on the Surface Geology of a Part of the Mississippi Valley, *Geol. Mag.*, Ser. 2, *6*: 412-420.

————, 1892. Comparative Chronology, *The American Anthropologist, 5:* 327-344.

————, 1893. Note on the Age of the Earth, *Science, 21* 309-310.

Macfarlane, Alexander, 1919. Lectures on Ten British Physicists of the Nineteenth Century (New York: Wiley).

Mackie, William, 1898-1905. The Saltiness of the Sea in Relation to the Geological Age of the Earth, *Trans. Edinburgh Geol. Soc., 8*: 240-255.

MacMillan, William D. On the Loss of Energy By Friction of the Tides. In Chamberlin, ed. (1909a), *Tidal and Other Problems,* pp. 69-75.

Macnair, Peter and Mort, Frederick, 1908. *History of the Geological Society of Glasgow; 1858-1908* (Glasgow: Geol. Soc.).

Malet, H. P., 1873. Earthquakes, *Geol. Mag., 10*: 74-80.

Mallet, Robert, 1880. Geology of the Henry Mountains, *Nature,* 22 July, *22*: 266.

Mandelbaum, Maurice, 1958. Darwin's Religious Views, *Jour. of the Hist. Ideas, 19*: 363-378.

Marchant, James, 1916. *Alfred Russel Wallace, Letters and Reminiscences,* 2 Vols. (London: Cassell).

Matthew, W. D., 1914. Time Ratios in the Evolution of Mammalian Phyla. A Contribution to the Problem of the Age of the Earth, *Science, 40*: 232-235.

Mayer, J. R., 1863a. On Celestial Dynamics, *Phil. Mag.,* Ser. 4, *25*: 241-248, 387-409, 417-428.

————, 1863b. Remarks on the Mechanical Equivalent of Heat, *Phil. Mag.,* Ser. 4, *25*: 493-522.

Merrill, George P., 1924. *The First One Hundred Years of American Geology* (New Haven; Yale Univ.).

Merz, John T., 1903-14. *A History of European Thought in the Nineteenth Century,* 4 Vols. (Edinburgh and London: Blackwood).

Meyer, Heinrich, 1951. *The Age of the World: A Chapter in the History of Enlightenment* (Allentown: multigraphed at Muhlenberg College).

Millhauser, Milton, 1959. *Just Before Darwin* (Middletown: Wesleyan Univ.).

Milne, J., 1880. Note upon the Cooling of the Earth, *Geol. Mag.,* 1880, Ser. 2, *7*: 99-102.

Mivart, St. George, 1871. *On the Genesis of Species* (New York: Appleton).

Morgan, C. Lloyd, 1878. Geological Time, *Geol. Mag.,* Ser. 2, *5*: 154-62, 199-207.

Moulton, Forest R., 1905. On the Evolution of the Solar System, *Astrophysical Jour., 22*: 165-81.

————, 1909a. On Certain Relations among the Possible Changes in the Motions of Mutually Attracting Spheres when Disturbed by Tidal Interactions. In Chamberlin, ed. (.1909a), *Tidal & Other Problems,* pp. 77-133.

————, 1909b. Notes on the Possibility of Fission of a Contracting Rotating Fluid Mass. In Chamberlin, ed. (1909a), *Tidal & Other*

Problems, pp. 135-160.

Munk, Walter H. & Macdonald, Gordon J. F., 1960. *The Rotation of the Earth* (Cambridge: Univ. Press).

Murphy, Joseph John, 1869. On the Nature and Cause of the Glacial Climate, *Geol. Mag., 6:* 331.

Murray, Robert H., 1925. *Science and Scientists in the Nineteenth Century* (London: Sheldon).

National Research Council, 1931. *Bulletin No. 80: Physics of the Earth-IV, The Age of the Earth* (Washington).

Newcomb, Simon, 1876. Review of Croll's Climate and Time with Especial Reference to the Physical Theories of Climate Maintained therein, *Am. Jour. Sci., 11:* 263-273.

———, 1878. *Popular Astronomy* (New York: Harper).

Newman, H. H., ed., 1926. *The Nature of the World and of Man* (Chicago: Univ. of Chicago).

Oldham, R. D., 1913. Radium and the Evolution of the Earth's Crust, *Nature,* 21 Aug., *91:* 635.

J. P., 1913. The Age of the Earth, *Nature,* 5 June, *91:* 343-44.

Parsons, S., 1877. The Nebular Hypothesis and Modern Genesis, *Methodist Quar. Rev.,* pp. 127-157.

Pearson, Karl, 1903. The Limits of Science, *Science,* 31 July, *18:* 140.

Perry, John, 1895a. On the Age of the Earth, *Nature,* 3 Jan., *51:* 224-227.

———, 1895b. On the Age of the Earth, *Nature,* 18 April, *51:* 582-85.

Petrie, W. M. Flinders, 1878. The Age of the Earth, *Nature,* 11 April, *17:* 465.

Phillips, John, 1837. *A Treatise on Geology,* 2 Vols. (London: Longman).

———, 1860a. Presidential Address, *Quar. Jour. Geol. Soc. Lon., 16:* xxx-lv.

———, 1860b. *Life on the Earth: Its Origin and Succession* (London: Macmillan).

———, 1864. Address to the Section of Geology at the Opening of the Thirty-fourth Meeting of the British Association, in Bath, September 15, 1864, *Geol. Mag., 1:* 175-180.

———, 1865. Presidential Address, *Brit. Assoc. Report* (Birmingham), pp. li-lxvii.

———, 1869. *Vesuvius* (Oxford: Clarendon).

Planck, Max, 1949. *Scientific Autobiography and Other Papers,* tr. Frank Gaynor (New York: Phil. Lib.).

Playfair, John, 1802. *Illustrations of the Huttonian Theory of the Earth* (Edinburgh: Cadell and Davies).

Plummer, John I., 1878. Age of the Sun in Relation to Evolution, *Nature*, 14 Feb., *17*: 303-04, and 7 Mar., *17*: 360.

Pouillet, Claude S. M., 1846. Memoir on Solar Heat, on the Radiating and Absorbing Powers of the Atmospheric Air, and On the Temperature of Space, *Taylor's Scientific Memoires*, 7 Vols. (London: R. & J. Taylor), IV: 44-90.

Poulton, Edward B., 1896. A Naturalist's Contribution to the Discussion Upon the Age of the Earth, *Brit. Assoc. Report* (Liverpool), pp. 808-828. Footnotes refer to reprint in E. B. Poulton, 1908. *Essays on Evolution* (Oxford: Clarendon), pp. 1-45.

Pratt, J. H., 1866. On the Level of the Sea During the Glacial Epoch in the Northern Hemisphere, *Phil. Mag.*, Ser. 4, *31*: 172-176.

Preston, S. Tolver, 1878. The Age of the Sun's Heat in Relation to Geological Evidence, *Nature*, 28 Mar., *17*: 423-24.

———, 1889. The Meteoric Theory of Nebulae, *Nature*, 7 Mar., *9*: 436-37 and 4 April, *39*: 535-36.

———, 1893. On the Supposed Conflict ·between Geology and Physics, *Geol. Mag.*, Ser. 3, *10*: 87-89.

Prestwich, Joseph, 1887. Considerations on the Date, Duration, and Conditions of the Glacial Period, with Reference to the Antiquity of Man, *Quar. Jour. Geol. Soc. Lon.*, *43*: 393-410.

———, 1895. *Collected Papers on Some Controverted Questions of Geology.* (London: Macmillan).

Ramsay, A. C., 1873-74. On the Comparative Value of Certain Geological Ages (Or Groups of Formations) Considered as Items of Geological Time, *Proc. Roy. Soc. Lon.*, *22*: 145-48.

Ramsay, William, 1904. The Source of Radium, *Nature*, 26 May, *70*: 80.

Ramsay, W. & Soddy, F., 1903a. Gases Occluded by Radium Bromide, *Nature*, 16 July, *68*: 246.

———, 1903b. Experiments in Radioactivity, and the Production of Helium from Radium, *Nature*, 13 Aug., *68*: 354-355.

Rappaport, Rhoda, 1964. Problems and Sources in the History of Geology, 1749-1810. *History of Science*, *3*: 60-78.

Reade, T. Mellard, 1877. On the Geological Significance of the Challenger Discoveries. In Reade (1879b) *Chemical Denudation*, pp. 32-34. Read to Liverpool Geol. Soc., 1877.

———, 1878a. Geological Time, *Proc. Liverpool Geol. Soc.*, *3*: 211-235. Also reprinted in Reade (1879b) *Chemical Denudation*, pp. 7-31. Read, 10 Oct. 1876.

————, 1878b. The Age of the World as Viewed by the Geologist and the Mathematician, *Geol. Mag.,* Ser. 2, *5*: 145-154.

————, 1879a. Limestone as an Index of Geological Time, In Reade (1879b) *Chemical Denudation,* pp. 43-61. Read to *Roy. Soc. Lon.,* 1879.

————, 1879b. *Chemical Denudation in Relation to Geological Time* (London: David Bogue).

————, 1883a. Examination of a Calculation of the Age of the Earth Based upon the Hypothesis of the Permanence of Oceans and Continents, *Geol. Mag.,* Ser. 2, *10*: 309-310.

————, 1883b. Reply to Mr. Wallace on the Age of the Earth, *Geol. Mag.,* Ser. 2, *10*: 571-73.

————, 1888. An Estimate of Post-Glacial Time, *Quar. Jour. Geol. Soc. Lon., 44*: 291-99.

————, 1893. Measurement of Geological Time, *Geol. Mag.,* Ser. 3, *10:* 97-100.

————, 1903. *The Evolution of Earth Structure* (London: Longman).

————, 1906. Radium and the Radial Shrinkage of the Earth, *Geol. Mag.,* Ser. 5, *3:* 79-80.

Reingold, Nathan, 1964. *Science in Nineteenth Century America, A Documentary History* (New York: Hill & Wang).

Roberts, R. D., 1880. Eccentricity and Glacial Epochs, *Geol. Mag.,* Ser. 2, 7: 143-44.

Rucker, Arthur W., 1901. Presidential Address, *Brit. Assoc. Report* (Glasgow), pp. 3-26.

Rudler, F. W., 1889-90. Experimental Geology, *Proc. Geologists' Assoc., 11*: 69-103.

Rudwick, Martin J. S., 1969a. Lyell on Etna, and the Antiquity of the Earth, *Toward a History of Geology,* C. Schneer, ed. (Cambridge: M.I.T.), pp. 288-304.

————, 1969b. The Glacial Theory, *History of Science, 8*: 136-57.

————, 1970. The Strategy of Lyell's Principles of Geology, *Isis, 61*: 5-33.

————, 1971. Uniformity and Progression: Reflections on the Structure of Geological Theory in the Age of Lyell, *Perspectives in the History of Science and Technology,* Duane H. D. Roller, ed. (Norman: Univ. of Oklahoma), pp. 209-227.

Russell, Alexander, 1912. *Lord Kelvin: His Life and Work* (London: Jack).

Russell, Henry Norris, 1921. A Superior Limit to the Age of the Earth's Crust, *Proc. Roy. Soc. Lon., 99*: 84-86.

Rutherford, Ernest (Baron Rutherford of Nelson), 1904a. The Radiation and Emanation of Radium, Pt. II, *Technics,* Aug., pp. 171-75. *Collected Papers,* I: 650-57.

————, 1904b. *Radio-Activity* (Cambridge: Univ. Press).

————, 1905a. Present Problems in Radioactivity, *Pop. Sci.* Monthly, *67*: 5-34.

————, 1905b. Les Problèmes Actuels de la Radioactivité, *Archives des Sciences Physiques et Naturelles, 19. Collected Papers,* I: 746-75.

————, 1905c. Radium—the Cause of the Earth's Heat, *Harper's Mag.,* Feb., pp. 390-96. *Collected Papers,.* I: 776-85.

————, 1906a. The Mass and Velocity of the α-Particles Expelled from Radium and Actinium, *Phil. Mag.,* Ser. 6, *12*: 348-378. *Collected Papers, I:* 880-900.

————, 1906b. *Radioactive Transformations* (New York: Scribners).

————, 1906c. The Recent Radium Controversy, *Nature,* 25 Oct., *74*: 634-35.

————, 1907. Some Cosmical Aspects of Radioactivity, *Jour. Roy. Ast. Soc. Canada,* pp. 145-165. *Collected Papers,* I: 917-31.

————, 1913. *Radioactive Substances and their Radiations* (Cambridge: Univ. Press).

————, 1915. Discussion on Radioactive Problems in Geology, *Brit. Assoc. Report* (Manchester), p. 432.

————, 1929. Origin of Actinium and Age of the Earth, *Nature,* 2 March, *123*: 313-14. *Collected Papers,* III: 216-17.

————, 1962-65. *The Collected Papers of Lord Rutherford of Nelson,* 3 Vols. (London: Allen & Unwin).

Rutherford, E. & Barnes, H. T., 1903. Heating Effect of the Radium Emanation, *Nature, 68*: 622 and *69*: 126. *Collected Papers,* I: 611-12.

————, 1904. Heating Effect of the Radium Emanation, *Phil. Mag.,* Ser. 6, 7: 200-219. *Collected Papers,* I: 625-39.

————, 1905. Heating Effect of the Rays from Radium, *Phil. Mag.,* Ser. 6, *9*: 621-28. *Collected Papers,* I: 793-98.

Rutherford, E. and Boltwood, B. B., 1905. The Relative Proportion of Radium and Uranium in Radio-active Minerals, *Am. Jour. Sci.,* Ser. 4, *20*: 55-56. *Collected Papers,* I: 801-02.

————, 1906. The Relative Proportion of Radium and Uranium in

Radio-Active Minerals, II, *Am. Jour. Sci.*, Ser. 4, *22*: 1-3. *Collected Papers*, I: 856-57.

Rutherford, E. and Geiger, Hans, 1908. The Charge and Nature of the α-Particle, *Proc. Roy. Soc. Lon.*, Ser. A, *81:* 162-73. *Collected Papers*, II: 109-120.

Rutherford, E. and Hahn, O., 1906. Mass of the α-Particles from Thorium, *Phil. Mag.*, Ser. 6, *12:* 371-78. *Collected Papers*, I: 901-06.

Rutherford, E. and Royds, T., 1908. The Nature of the α-Particle, *Memoirs of Man. Lit. & Phil.*, Ser. 4, *53*: 1-3. *Collected Papers*, II: 134-135.

Rutherford, E. and Soddy, F., 1903. Radioactive Change, *Phil. Mag.*, Ser. 6, *5*: 576-91. *Collected Papers*, I: 596-608.

Salisbury, Robert Cecil, 3rd Marquis, 1894. Presidential Address, *Brit. Assoc. Report* (Oxford), pp. 3-15.

Schaeberle, John M., 1907a. The Effective Surface-Temperature of the Sun and the Absolute Temperature of Space, *Science, 26*: 718-19.

————, 1907b. The Probable Origin and Physical Structure of Our Siderial and Solar Systems, *Science, 26*: 877-78.

————, 1908a. Geological Climates, *Science, 27*: 894.

————, 1908b. The Earth as a Heat-Radiating Planet, *Science, 27*: 392-93.

————, 1908c. On the Origin and Age of the Sedimentary Rocks, *Science, 28*: 562-565.

Schiller, F. C. S., 1913. Radioactivity and the Age of the Earth, *Nature*, 26 June, *91*: 424 and 17 July, *91*: 505.

Schneer, Cecil J., ed., 1969. *Toward A History of Geology* (Cambridge: M.I.T.).

Schuchert, Charles, 1923. The Earth's Changing Surface and Climate During Geological Time, *The Evolution of the Earth and Its Inhabitants*, G. A. Baitsell, ed. (New Haven: Yale Univ.), pp. 50-80.

————, 1931. Geochronology, or The Age of the Earth on the Basis of Sediments and Life. In National Research Council (1931), *Bulletin No. 80*, pp. 10-64.

Schvarcz, Julius, 1868. On the Internal Heat of the Earth, *Geol. Mag., 5:* 26.

Scrope, G. Poulett, 1827. *Memoir on the Geology of Central France; Including the Volcanic Formations of Auvergne, the Velay and the Vivarois*, 2 Vols. (London: Longman).

————, 1868. Some Observations on the Supposed Internal Fluidity of the Earth, *Geol. Mag., 5*: 537-39.

————, 1869. On the Supposed Internal Fluidity of the Earth, *Geol. Mag., 6*: 145-47.

————, 1873. On 'Blocky' Rock Surfaces, and the Theory of the Shrinking Nucleus of the Globe, *Geol. Mag., 10*: 291-95.

See, T. J. J., 1899. An Extension of Helmholtz's Theory of the Heat of the Sun, *Science,* 26 May, *9*: 737-740.

————, 1905. Current Theories of the Consolidation of the Earth, *Nature,* 11 May, *72*: 30-31.

————, 1907. The Age of the Earth's Consolidation, *Scientific American* (Suppl.), *64*: 366.

Seeley, H. G., 1909. *The Story of the Earth in Past Ages* (New York: Univ. Society).

Shairp, John Campbell, Tait, P. G., and Adams-Reilly, A., 1873. *Life and Letters of James David Forbes, F.R.S.* (London: Macmillan).

Shaler, N. S., 1868. On the Formation of Mountain Chains, *Geol. Mag., 5*: 511-17.

Shelton, H. S., 1911. Modern Theories of Geological Time, *Contemporary Review, 99*: 195-209.

————, 1913-14a. The Age of the Earth, *Science Progress, 8*: 168-170.

————, 1913-14b. Some Aspects of Geologic Time, *Science Progress, 8*: 250-274.

————, 1913-14c. A Suggestion Concerning the Origin of Radioactive Matter, *Science Progress, 8*: 456-459.

Siemens, C. William, 1882. On the Conservation of Solar Energy, *Proc. Roy. Soc. Lon., 33*: 389-98.

Slaughter, William B., 1876. *The Modern Genesis* (New York: Nelson and Phillips).

Slosson, Edwin E., 1923. Geologists get an Extension of Time, *Scientific Monthly, 16*: 329-31.

Soddy, Frederick, 1904a. *Radio-Activity, An Elementary Treatise from the Standpoint of the Disintegration Theory* (London: "The Electrician").

————, 1904b. The Life History of Radium, *Nature,* 12 May, *70*: 30.

————, 1905a. The Origin of Radium, *Nature,* 26 Jan., *71*: 294.

————, 1905b. The Production of Radium from Uranium, *Phil. Mag.,* Ser. 6, *9*: 768-79.

————, 1906. The Recent Controversy on Radium, *Nature,* 20 Sept., *74*: 516-18.

————, 1909. The Production of Radium from Uranium, *Nature*, 13 May, *80*: 303-304.

————, 1913. The Origin of Actinium, *Nature*, 21 Aug., *91*: 634-35.

————, 1915a. The Density of Lead from Ceylon Thorite, *Nature*, 4 Feb., *94*: 615.

————, 1915b. Discussion on Radio-Active Problems in Geology, *Brit. Assoc. Report* (Manchester), p. 434.

————, 1917. The Atomic Weight of 'Thorium' Lead, *Nature*, Feb., *98*: 469.

Sollas, William J., 1877. On Evolution in Geology, *Geol. Mag.*, Ser. 2, *4*: 1-7.

————, 1895. The Age of the Earth, *Nature*, 4 April, *51*: 533-34.

————, 1900. Evolutional Geology, *Smithsonian Report*, pp. 289-315.

————, 1901. Age of the Earth, *Current Literature*, *30*: 579-80.

————, [1905]. *The Age of the Earth and Other Geological Studies* (London: Unwin).

————, 1909. The Anniversary Address of the President, *Quar. Jour. Geol. Soc. Lon.*, *65*: 1-cxxii.

————, 1921. The Age of the Earth, *Nature*, 27 Oct., *108*: 281-283.

Stevenson, Robert Louis, 1887. Memoir of Fleeming Jenkin, *Papers of Fleeming Jenkin*, S. Colvin and J. A. Ewing, eds., 2 Vols. (London: Longman).

Stewart, Balfour, 1896. *The Conservation of Energy* (New York: Appleton).

Stromeyer, C. E., 1915. The Age of the Earth, *Nature*, 6 May, *95*: 259-60.

Strutt, Robert J. (4th Baron Rayleigh), 1903. Radium and the Sun's Heat, *Nature*, 15 Oct., *68*: 572.

————, 1904a. *Becquerel Rays and the Properties of Radium* (London: Edward Arnold).

————, 1904b. A Study of the Radioactivity of Certain Minerals and Mineral Waters, *Nature*, 17 Mar., *69*: 473-75.

————, 1904c. The Occurrence of Radium with Uranium, *Nature*, 7 July, *70*: 222.

————, 1905a. The Rate of Formation of Radium, *Nature*, 17 Aug., *72*: 365.

————, 1905b. Radioactivity of Ordinary Matter in Connection with the Earth's Internal Heat, *Nature*, 21 Dec., *73*: 173.

————, 1905c. On Radioactive Minerals, *Proc. Roy. Soc. Lon.*, Ser. A, *76*: 88-101.

————, 1906. On the Distribution of Radium in the Earth's Crust, and On the Earth's Internal Heat, *Proc. Roy. Soc. Lon.*, Ser. A, 77: 472-85.

————, 1908a. Note on the Association of Helium and Thorium in Minerals, *Proc. Roy. Soc. Lon.*, Ser. A, *80*: 56-57.

————, 1908b. Helium and Radio-Activity in Rare and Common Minerals, *Proc. Roy. Soc. Lon.*, Ser. A, *80*: 572-94.

————, 1908c. The Leakage of Helium from Radio-active Minerals, *Proc. Roy. Soc. Lon.*, Ser. A, *82*: 166-69.

————, 1908d. Radium and the Earth's Heat, *Nature*, 20 Feb., 77: 365-66.

————, 1908e. Radio-active Changes in the Earth, *Nature*, 17 Dec., 79: 206-208.

————, 1908f. On the Accumulation of Helium in Geological Time, *Proc. Roy. Soc. Lon.*, Ser. A, *81*: 272-79.

————, 1909a. On the Accumulation of Helium in Geological Time, Pt. II, *Proc. Roy. Soc. Lon.*, Ser. A, *83*: 96-99.

————, 1909b. On the Accumulation of Helium in Geological Time, Pt. III, *Proc. Roy. Soc. Lon.*, Ser. A, *83*: 298-301.

————, 1909c. A Direct Estimate of the Minimum Age of Thorianite, *Nature*, 13 May, *80*: 303.

————, 1910a. On the Accumulation of Helium in Geological Time, Pt. IV, *Proc. Roy. Soc. Lon.*, Ser. A, *84*: 194-96.

————, 1910b. Measurements of the Rate at which Helium is Produced in Thorianite and Pitchblende, With a Minimum Estimate of Their Antiquity, *Proc. Roy. Soc. Lon.*, Ser. A, *84*: 379-388.

————, 1921. The Age of the Earth, *Nature*, 27 Oct., *108*: 279-81.

Tait, Peter Guthrie, 1869. Geological Time, *North British Review*, July, *50*: 215-233.

————, 1876. *Lectures on Some Recent Advances in Physical Science* (London: Macmillan).

————, 1895. On the Age of the Earth, *Nature*, 3 Jan., *51*: 226.

Taylor, Andrew, 1869-74. On the Probability of Our Successfully Calculating the Antiquity of the Earth, *Trans. Edin. Geol. Soc.*, *2*: 402-411.

Thiselton-Dyer, W. T., 1903. The Limits of Science, *Science*, 31 July, *18*: 138-39, 141-43.

Thompson, Silvanus P., 1908. The Kelvin Lecture: The Life and Work of Lord Kelvin, *Jour. of the Institution of Electrical Engineers, 41*: 401-423.

————, 1910. *The Life of William Thompson, Baron Kelvin of Largs,* 2 Vols. (London: Macmillan).

Thomson, John Joseph, 1936. *Recollections and Reflections* (London: Bell).

Toulmin, Stephen & Goodfield, June, 1966. *The Discovery of Time* (New York: Harper).

Twain, Mark, 1962. *Letters from the Earth,* Bernard DeVoto, ed. (New York: Harper).

Tylor, Alfred, 1853. On Changes of the Sea Level Effected by Existing Physical Causes during Stated Periods of Time, *Phil. Mag.,* Ser. 4, *5*: 258-81.

Tyndall, John, 1874. Presidential Address, *Brit. Assoc. Report* (Belfast), pp. lxvi-xcvii.

Upham, Warren, 1893. Estimates of Geologic Time, *Am. Jour. Sci., 145*: 209-220.

————, 1906. Geological Time, *Pop. Astronomy, 14*: 264-276.

Walcott, Charles D., 1893. Geologic Time, As Indicated by the Sedimentary Rocks of North America, *American Geologist,* Dec., *12*: 342-68.

Wallace, Alfred Russel, 1869a. Geological Climates and the Origin of the Species, *Quar. Review, 126*: 359-394.

————, 1869b. The Origin of Species Controversy, *Nature,* 25 Nov., *1*: 105-07 & 132-33.

————, 1870. The Measurement of Geological Time, *Nature,* 17 Feb., *1*: 399-401 & 452-455.

————, 1880a. *Island Life: or the Phenomena and Causes of Insular Faunas and Floras, Including a Revision and Attempted Solution of the Problem of Geological Climates* (London: Macmillan).

————, 1880b. Dr. Croll's Eccentricity Theory, *Geol. Mag.,* Ser. 2, *7*: 284-85.

————, 1883. Reply to T. Mellard Reade on the Age of the Earth, *Geol. Mag.,* Ser. 2, *10*: 478-80.

————, 1892-93. The Earth's Age, *Nature,* 22 Dec., *47*: 175, and 5 Jan., *47*: 227.

————, 1895a. The Age of the Earth, *Nature,* 25 April, *51*: 607.

————, 1895b. Uniformitarianism in Geology, *Nature,* 2 May, *52*: 4.

————, 1905. *My Life,* 2 Vols. (London: Chapman & Hall).

Ward, J. Clifton, 1868. Internal Fluidity of the Earth, *Geol. Mag., 5*: 581-82.

————, 1869. Suggestions as to Geological Time, *Geol. Mag., 6*: 8-13.

Waterston, John James, 1853. On a Law of Mutual Dependence Between Temperature and Mechanical Force, *Brit. Assoc. Report* (Hull), pp. 11-12.

Webster, Arthur Gordon, 1908. Lord Kelvin, *Science,* 3 Jan., *27*: 1-8.

Whetham, W. C. D., 1904. The Life History of Radium, *Nature,* 5 May, *70*: 5.

————, 1905. The Origin, of Radium, *Nature,* 2 Feb., *71*: 319.

Whittaker, Thomas, 1891. The Philosophical Basis of Evolution by James Croll, *Mind, 16*: 268-271.

Wilkins, Thurman, 1958. *Clarence King* (New York: Macmillan).

Williams, Henry Shaler, 1893a. The Making of the Geological Time Scale, *Jour. of Geol., 1*: 180-197.

————, 1893b. The Elements of the Geological Time Scale, *Jour. of Geol., 1*: 283-295.

Williams, Henry Smith, 1904. *A History of Science,* 5 Vols. (New York: Harper).

Williams, William Mattieu, 1870. *The Fuel of the Sun* (London).

————, 1882. Current Discussions in Science, *The Humboldt Library of Popular Science* (New York: Humboldt), IV: 207-219.

Wilson, Harold A., 1908. Radium and the Earth's Heat, *Nature,* 20 Feb., *77*: 365.

Wilson, J. M., 1870. Sir W. Thomson and Geological Time, *Nature,* 14 April, *1*: 606.

Wilson, Leonard G., 1964. The Development of the Concept of Uniformitarianism in the Mind of Charles Lyell, *Actes X₊ Cong. Intern. Hist. Sci.* (Paris), II: 993-96.

————, 1967. The Origins of Charles Lyell's Uniformitarianism, *Geol. Soc. Am.,* Special Paper 89, pp. 35-62.

————, 1969. The Intellectual Background to Charles Lyell's *Principles of Geology,* 1830-33, *Toward a History of Geology,* C. Schneer, ed. (Cambridge: M.I.T.), pp. 426-43.

————, 1971. Sir Charles Lyell and the Species Question, *American Scientist, 59*: 43-55.

Wilson, W. E., 1903. Radium and Solar Energy, *Nature,* 9 July, *68*: 222.

Winchell, Alexander, 1870. *Sketches of Creation* (New York: Harper).

————, 1876. Climate and Time, *The International Review, 3*: 519-26.

————, 1883. *World Life, or Comparative Geology* (Chicago: Griggs).

————, 1889. Views on Prenebular Conditions, *American Geologist, 4*: 196-205.

Winnik, Herbert C., 1970. Science and Morality in Thomas C. Chamberlin, *Jour. Hist. Ideas, 31*: 441-56.

Woeikof, A., 1886. Examination of Dr. Croll's Hypothesis of Geological Climates, *Am. Jour. Sci., 131*: 161-178.

Wood, Searles V., 1876. The Climate Controversy, *Geol. Mag.*, Ser. 2, *3*: 385-98, 442-51.

————, 1883. On the Cause of the Glacial Period, *Geol. Mag.*, Ser. 2, *10*: 293-302.

Woodward, Henry, 1873. Geologists' Association, University College, London. Address at the Opening Session, November 7th, 1873, *Geol. Mag., 10*: 529-552.

Woodward, Horace B., 1911. *History of Geology* (London: Watts).

Woodward, Robert Simpson, 1889. The Mathematical Theories of the Earth, *Proc. Am. Assoc. Adv. Sci., 38*: 49-69.

Zeuner, Frederick E., 1946. *Dating The Past: An Introduction to Geochronology* (London: Methuen).

Zittel, Karl A. von, 1901. *History of Geology and Palaeontology*, Maria M. Ogilvie-Gordon, tr. (London: Walter Scott).

Bibliography of Works Published Since 1974

Albritton, Claude C., Jr. *The Abyss of Time* (San Francisco: Freeman, 1980).

Bartholomew, Michael. The Non-progress of Non-progression: Two Responses to Lyell's Doctrine, *Brit. Jour. for the Hist. of Sci.*, 1976, *9*: 166–74.

—————— The Singularity of Lyell, *Hist. of Sci.*, 1979, *17*: 276–93.

Bowler, Peter J. *The Eclipse of Darwinism* (Baltimore: Johns Hopkins U. Press, 1983).

—————— *The Non-Darwinian Revolution: Reinterpreting a Historical Myth* (Baltimore: Johns Hopkins U. Press, 1988).

Brush, Stephen G. Thermodynamics and History, *The Grad. Jour.*, Spring 1967, 477–565.

—————— *The Temperature of History* (New York: Burt Franklin, 1978).

—————— Planetary Science: From Underground to Underdog, *Scientia*, 1978, *132*: 771–87.

—————— A Geologist among Astronomers: the Rise and Fall of the Chamberlin-Moulton Cosmogony, *Jour. for the Hist. of Astron.*, 1978, *9*: 1–41, 77–104.

—————— Nineteenth-Century Debates about the Inside of the Earth: Solid, Liquid, or Gas? *Annals of Sci.*, 1979, *36*: 225–54.

—————— Discovery of the Earth's Core, *Amer. Jour. of Physics*, 1980, *48*: 705–24.

———— Finding the Age of the Earth by Physics or by Faith, *Jour. of Geol. Educ.*, 1982, *30*: 34–58.

———— Whole Earth History, *Hist. Studies in Phys. and Biol. Sciences*, 1987, *17*: 352–53.

———— The Nebular Hypothesis and the Evolutionary Worldview, *Hist. of Sci.*, 1987, *25*: 245–78.

———— The Age of the Earth in the Twentieth Century, *Earth Sci. Hist.*, 1989, *8*: 170–82.

Cannon, Susan Faye. *Science in Culture: The Early Victorian Period* (New York: Science History Publications, 1978).

Cannon, W. Faye. Charles Lyell, Radical Actualism, and Theory, *Brit. Jour. for the Hist. of Sci.*, 1976, *9*: 104–20.

Dean, Dennis R. The Age of the Earth Controversy: Beginnings to Hutton, *Annals of Sci.*, 1981, *38*: 435–56.

———— Through Science to Despair: Geology and the Victorians, *Annals of the N.Y. Acad. of Sci.*, 1981, *360*: 111–33.

Di Gregorio, Mario A. *T. H. Huxley's Place in Natural Science* (New Haven: Yale U. Press, 1984).

Faul, Henry. A History of Geologic Time, *Amer. Scientist*, 1978, *66*: 159–65.

Gillmor, C. Stewart. The Place of the Geophysical Sciences in Nineteenth Century Natural Philosophy, *EOS*, 1975, *56*: 4–7.

Gould, Stephen J. False Premise, Good Science, *Nat. Hist.*, 1983, *92*: 20–26. Reprinted in *The Flamingo's Smile* (New York: Norton, 1985), pp. 126–38.

Greene, Mott T. *Geology in the Nineteenth Century* (Ithaca: Cornell U. Press, 1982).

Hallam, A. *Great Geological Controversies* (Oxford: Oxford U. Press, 1983).

Hamilton, B. M. British Geologists' Changing Perceptions of Precambrian Time in the Nineteenth Century, *Earth Sci. Hist.*, 1989, *8*: 141–49.

Hamlin, Christopher. James Geikie, James Croll, and the Eventful Ice Age, *Annals of Sci.*, 1982, *39*: 565–83.

Harrison, T. M. Comment on "Kelvin and the Age of the Earth," *Jour. of Geol.*, 1987, *95*: 725–27.

Hufbauer, Karl. Astronomers Take Up the Stellar-Energy Problem, 1917–1920, *Hist. Studies in Phys. Sci,*, 1981, *11*: 277–303.

James, Frank A. J. L. Thermodynamics and Sources of Solar Heat, 1846–1862, *Brit. Jour. for Hist. of Sci.*, 1982, *15*: 155–81.

Laudan, Rachel. *From Mineralogy to Geology: The Foundations of a Science, 1650–1830* (Chicago: U. of Chicago Press, 1987).

Lawrence, Philip. Charles Lyell versus the Theory of Central Heat: a Reappraisal of Lyell's Place in the History of Geology, *Jour. of the Hist. of Biol.*, 1978, *11*: 101–28.

Moore, James R. *The Post-Darwinian Controversies* (Cambridge: Cambridge U. Press, 1979).

Nahin, P. J. Oliver Heaviside, Fractional Operators, and the Age of the Earth, *IEEE Transactions on Educ.*, 1985, *28*: 94–104.

Ospovat, Dov. Lyell's Theory of Climate, *Jour. of the Hist. of Biol.*, 1977, *10*: 317–39.

Porter, Roy. Charles Lyell and the Principles of the History of Geology, *Brit. Jour. for the Hist. of Sci.*, 1976, *9*: 91–103.

———— *The Making of Geology* (Cambridge: Cambridge U. Press, 1977).

———— Gentlemen and Geology: the Emergence of a Scientific Career, 1660–1920, *Hist. Jour.*, 1978, *21*: 809–36.

Richter, F. N. Kelvin and the Age of the Earth, *Jour. of Geol.*, 1986, *94*: 395–401.

Rudwick, Martin J. S. Poulett Scrope on the Volcanoes of Auvergne: Lyellian Time and Political Economy, *Brit. Jour. for the Hist. of Sci.*, 1974, *7*: 205–42.

———— Caricature as a Source for the History of Science: De la Beche's Anti-Lyellian Sketches of 1831, *Isis*, 1975, *66*: 534–60.

———— Charles Lyell's Dream of a Statistical Palaeontology, *Palaeontology*, 1978, *21*: 225–44.

———— *The Great Devonian Controversy* (Chicago: U. of Chicago Press, 1985).

Ruse, Michael. *The Darwinian Revolution* (Chicago: U. of Chicago Press, 1979).

Secord, James A. *Controversy in Victorian Geology: The Cambrian-Silurian Dispute* (Princeton: Princeton U. Press, 1986).

Servos, John W. Mathematics and the Physical Sciences in America, 1880–1930, *Isis*, 1986, *77*: 611–29.

Sharlin, Harold I. *Lord Kelvin: the Dynamic Victorian* (University Park: Pennsylvania State U. Press, 1979).

Shea, James H. Twelve Fallacies of Uniformitarianism, *Geol.*, 1982, *10*: 455–60.

Smith, Crosbie W. William Thomson and the Creation of Thermodynamics: 1840–1855, *Archive for Hist. of the Exact Sciences*, 1976, *16*: 231–88.

_____ William Hopkins and the Shaping of Dynamical Geology: 1830–1860, *Brit. Jour. for the Hist. of Sci.*, 1989, *22*: 27.

Smith, Crosbie, & Wise, M. Norton. *Energy and Empire: A Biographical Study of Lord Kelvin* (Cambridge: Cambridge U. Press, 1989).

Wilson, David B. *Kelvin and Stokes: A Comparative Study in Victorian Physics* (Bristol: Adam Hilger, 1987).

Wilson, Leonard G. *Charles Lyell, the Years to 1841: The Revolution in Geology* (New Haven: Yale U. Press, 1972).

Wood, Robert Muir. *The Dark Side of the Earth* (London: George Allen & Unwin, 1985).

Yeo, R. Science and Intellectual Authority in Mid-Nineteenth-Century Britain, *Victorian Studies*, 1984, *28*: 5–31.

Yochelson, Ellis L. "Geologic Time" as Calculated by C. D. Walcott, *Earth Sci. Hist.*, 1989, *8*: 150–58.

Index